WHY THINGS BREAK

WHY THINGS BREAK

UNDERSTANDING THE WORLD BY THE WAY IT COMES APART

MARK E. EBERHART

THREE RIVERS PRESS • NEW YORK

Published by Three Rivers Press, New York, New York.
Member of the Crown Publishing Group, a division of Random House, Inc.
www.crownpublishing.com

THREE RIVERS PRESS and the Tugboat design are registered trademarks of Random House, Inc.

Originally published in hardcover by Harmony Books, a division of Random House, Inc., in 2003.

Printed in the United States of America

Design by Barbara Sturman

Library of Congress Cataloging-in-Publication Data
Eberhart, Mark.
Why things break : understanding the world by the way it comes apart / Mark Eberhart.
1. Fracture mechanics. I. Title.
TA409.E32 2003
620.1'12—dc21 2003004376

ISBN 1-4000-4883-4

10 9 8 7

First Paperback Edition

To Cheryl,
who bends but never breaks

CONTENTS

ACKNOWLEDGMENTS

Though mine is the story of a personal journey along seldom-traveled back roads of science, it would not have been possible without the help of so many. Among those to whom I am most indebted are Greg Olson, who has been a constant source of inspiration; Keith Johnson, for providing me with an abundance of opportunity; Ron Latanision, who encouraged me when I doubted myself; Pat Martin, who introduced me to the people who actually make things that break; Dimitri Vvedensky, who taught me that it only takes a little longer to do things right; and James MacLaren, for providing me with the tools to do it.

THERE WAS A GAMBLER who loved to play the ponies. He studied the horses religiously, believing that if he knew a bit more and was just a bit smarter he would be able to come up with a winning system. Unfortunately the problem was just too complex and after years of study the best the gambler could do was win 55 percent of the time. In desperation the gambler decided to seek out the smartest person he could find—an eminent physicist at a local university. After explaining the problem to the scientist, the physicist agreed that this was indeed a difficult problem and instructed the gambler to come back in a week. When the gambler returned the physicist gave him a list of the horses he predicted would win each of that day's races. The gambler bet everything he owned on the scientist's selections only to lose it all. Enraged, the gambler confronted the physicist, demanding to know how the selections were made. "Well," the physicist said, "I began by assuming a spherical horse."

WHY THINGS BREAK

ATOMS, MARBLES, AND FRACTURE

CHAPTER

1

What incredible luck. The waitress had just unknowingly placed the most amazing water glass on our table. Halfway up the glass was a crack about two centimeters long. This was one of those fantastic cracks where neither end intersected a surface. These are stable and, if left alone, will simply hibernate. Water does not leak from these cracks, their presence is known only by the reflection of light from their surfaces. If disturbed, however, they wake up, sometimes violently, growing with incredible speed, often branching as they go, reducing whatever contained them in their quiescent state to a pile of razor-sharp shards.

Though the crack in this water glass was a rare find, it was even more remarkable in that it was oriented nearly parallel to the bottom of the glass. If gently awakened, the ends of this crack could be made to grow around the glass and meet at the same point, dividing the glass into two parts. Quickly downing the water, I used the handle of a butter knife to tap on the glass, ever so gently, near the tips of the crack. Too sharp a blow and the crack would become uncontrollable. With each tap, the crack grew slightly and stopped. Slowly the ends of the crack worked their way around the glass and, with no apparent sound, they joined. As if by magic, aided only by the butter knife "wand," the glass had been separated.

I was delighted with my carefully divided glass. My lunch companions, however, were less than pleased. I was, after all, with my impressionable young nieces and their parents. The looks on their faces suggested that I had just committed the most ill-conceived of social faux pas. Though this incident occurred nearly ten years ago, the breaking of the water glass is still a subject that causes my nieces and their parents to reflect on my integrity. My "crime" was a minor one. Indeed, the glass would not have survived even one more washing. The thermal strains caused by heating and cooling would have marked the end of the glass's useful life, and such a marvelous crack deserved a more significant death.

My interest in cracks and fracture began in early childhood, when I became fascinated by the idea that it might be possible to *prevent* things from breaking. I imagined what the world would be like if things never broke. In my child's mind, I pictured both the great and ordinary creations of humankind surviving the ages untouched and pristine. Little did I realize that this simple fantasy would direct my life and open doors I never thought existed.

Perhaps my interest was a by-product of growing up in the 1950s and 1960s. Every child lived with the fear that "the bomb" could be dropped at any minute. In school, it was common for a teacher to open the door of a classroom and yell, "Duck and cover!" In response to this warning, we students were expected to fling ourselves to the floor in a modified fetal position, with hands clasped across the back of the neck. In the absence of the requisite warning, the duck-and-cover position was to be assumed when we saw the blinding flash of an atomic-bomb explosion. I firmly believed that the duck-and-cover position would protect me from an atomic blast, but I knew the inanimate part of the world would surely be destroyed. After all, the purpose of duck-and-cover was to protect us from the flying pieces of objects broken by the blast. This seemed to be such an incredible waste. How could people work so hard to build things, only to see them destroyed? To me, making things that didn't break was one way around the destruction that nuclear war would bring.

The fear of pending nuclear annihilation may have seeded my interest in combating fracture, but the same fear also directed me down a path that would ultimately provide the tools necessary to achieve that goal. An axiom of the time was that "the power of the atomic bomb was unleashed by splitting the atom." This concerned me. If splitting a single atom were to cause an atomic explosion, was it possible that someone might inadvertently slice through one while using a knife or a pair of scissors? I pictured little mushroom clouds over thousands of dinner tables, each the result of an accident with a butter knife, but this never happened. Fortunately, the expected news story—"Today the John and Betty Smith family, their home, and the surrounding neighborhoods were demolished as John attempted to butter his bread"—never made the six-o'clock news. The only explanation for the absence of unintended nuclear blasts was that a butter knife was incapable of slicing through an atom. As a six-year-old, I began to construct a model that would explain this observation.

I envisioned the atoms of the butter as marbles spread out on the floor so that they just touched. Because everything was made from atoms, the edge of a knife could also be pictured as marbles, perhaps marbles of different sizes, peewees or boulders, but still marbles. The act of cutting the butter was like dragging the "knife" marbles through the other marbles on the floor. In my mind's eye, I picture holding a marble and pulling it through the marbles representing the butter. The knife would separate the marbles into two groups, but never would another marble be cut in half.

Though I slept easier knowing that making breakfast was unlikely to trigger Armageddon, a new question began to preoccupy me. There had to be *something* that held atoms together. If atoms were like marbles, they would just puddle out when taken out of their container. The knife was cutting not the marbles themselves, but whatever it was that held the atoms together.

I can't remember actually performing the marble experiment, and I doubt that I ever did; I valued my marbles too much to actually use them. Those of my friends who actually played the game had the most pitted and ugly marbles you could imagine. With only a few exceptions, my marbles remained as perfect as the day they were purchased. To me, a chipped one was worthless, having lost its value with its beauty. The exceptions were those marbles that were intentionally fractured to make them even more beautiful. The procedure is simple. Place a marble (a "cleary" is best; that's a marble made from a single piece of colored glass, devoid of internal decoration) on a cookie sheet and heat in an oven to 250° F. Remove the marble when heated, and immediately drop it into cold water. Under these conditions, most marbles will respond by producing an array of internal fractures. The reflection of light from these internal surfaces produces an esthetically pleasing effect.

The problem with such marbles is that almost any blow will cause the cracks to run, leaving you with a pile of broken glass. They are useless from a utilitarian viewpoint; they can't even be carried in a marble bag for fear of shattering them. So, though I had come up with a method to preserve, and even extend, the beauty of my marbles, it was not a practical solution, since it required that they not be used. There had to be another way. Was it possible to make a glass marble that would not pit when used?

Having already developed an idea about what happens

when something is cut, it took only a tiny step to picture what happened when something broke. Once again, I pictured the atoms of the glass as marbles packed together on the floor. This time a marble was shot at the pile, just as in the real game, dislodging other marbles from the central group. These dislodged marbles I thought of as the atoms of the broken chips of glass. If one wanted to make glass that would not chip, then whatever held the atoms of the glass together must be made stronger.

I still had no idea what held those atoms together. I had several small magnets, however, and I imagined the force holding the magnets together had to be similar to that holding atoms together. The problem with magnets, however, is their shape. Mine were horseshoe magnets and didn't look much like spherical atoms. Despite all my efforts, I could not seem to locate magnetic marbles. It appeared that my very first scientific investigation had come to a grinding halt at the ripe old age of six. Three years would pass before it could be revived.

In third-grade science class, we were observing magnetic fields by placing a sheet of paper over a magnet and then sprinkling iron filings on the paper. The purpose of the experiment was to observe how the filings lined up in the magnetic field. It was neat that something invisible could be made visible so easily. Even neater, however, was the fact that the iron filings became magnetic and attracted each other. Would little iron marbles behave the same way?

At the hardware store, they told me little iron marbles were called ball bearings and they came in many sizes. With my birthday money, I bought about a hundred BB-sized ball bearings and two bar magnets. At home, I set the bar magnets on end underneath a piece of cardboard and then poured the ball bearings on top. It worked exactly as it was supposed to; the ball bearings became magnetic and attracted each other. By gently tapping the cardboard, the bearings would arrange themselves in a periodic array. If the two magnets were set sufficiently far apart, two arrays could be made. Then, by dragging the magnets underneath the cardboard, the groups of bearings came together to form a single island. Depending on how the magnets were arranged (the north end up on one and the south end up on the other, or both magnets with the same end up), different results were obtained. Sometimes the groups of bearings would form a single array in which it was impossible to distinguish to which group a ball bearing originally belonged, and sometimes they would coalesce into a single group of bearings with an odd line that marked the boundary between the two. If the cardboard was gently vibrated, the boundary would disappear. By reversing the process, the island could be made to fall apart in two pieces. The boundary between the two islands of bearings looked different depending on how fast the magnets were pulled apart.

I played with my magnets and ball bearings for hours and discovered that by changing the strength of the magnet, the forces holding the ball bearings together could also be changed. This was accomplished by simply moving the magnet farther away from the cardboard. I made little spacers that fit between the magnet and the cardboard for this purpose. The larger the spacer, the weaker the bond. Using a strong magnet with no spacer, I would shape little wedges of ball bearings and ram them into the flat surfaces of oblong shapes, held together with a magnet separated from the cardboard by a large spacer. The oblong shape would deform and remain deformed when the wedge was withdrawn. On occasion, some of the bearings of the oblong were pulled away by the wedge.

After all the play, I was convinced that the different behaviors of ball-bearing atoms resulted from the strength of the forces that held them together. But I still did not know what held real atoms together. I was determined to find out.

Seven years later, in a high school chemistry class, I finally had my answer. They were *chemical* bonds. It turned out that the science of chemistry was concerned almost entirely with the study of these bonds—moving them, strengthening them, and so on. If you wanted to do something to a bond, you had to be a chemist, and that is exactly what I intended to become.

Though I now knew the things holding atoms together were called chemical bonds, actually understanding how those bonds worked would require considerable effort. Unlike baseballs, cars, and magnets, which respond to force according to the laws discovered by Newton, bonds respond in a very different way. So different and strange is their response that it was not until 1926 that the laws governing their behavior were discovered. Things that behave according to the laws of Newton are said to behave classically and obey the laws of classical mechanics. Bonds, however, obey the laws of quantum mechanics.

From the perspective of a high school student, the laws

of quantum mechanics appear to make no sense. For example, those laws allow something to be in two places at the same time. Though this violates common sense, common sense is based on experience, and our experience is consistent with the laws of classical mechanics. Understanding quantum mechanics requires everyday experience to be set aside, and in its place one substitutes the mathematical expressions describing the new laws discovered in 1926 by Erwin Schrödinger. Unfortunately, as a high school student I did not have the mathematical maturity necessary to understand quantum mechanics.

The desire to have my marbles and use them too had made the agenda clear. Study chemistry and learn how to manipulate bonds, study mathematics to understand quantum mechanics, and study quantum mechanics and learn how bonds worked. I elected to pursue this agenda at the University of Colorado in Boulder.

A student living in Boulder is faced with a number of distractions. Boulder is located at the base of the Rocky Mountains. Some of the most spectacular rock climbing in the world can be found just a few minutes from campus. A little longer drive away is world-class skiing. Of course, the ski season begins in December and generally ends in April, requiring skiers to find some other activity for the summer months. Whitewater kayaking was my summer activity. Chemistry and mathematics classes, skiing in the winter and spring, kayaking in the summer and fall, left little time to do much else, like get a job. Little did I suspect that my life as an unemployed student ski-and-kayak bum would transform me into a real expert on why things break.

As I studied chemistry, I constantly anticipated that the answer to why things broke was just around the corner. With the beginning of each semester came the same ritual. I would search through my newly purchased textbooks looking for some reference to fracture, some explanation. There were literally full chapters discussing the techniques for manipulating chemical bonds and transforming one molecule into another, but nothing about why those bonds unsurprisingly broke. The study of quantum mechanics was more involved with the interaction of molecules and solids with light than with mechanical forces. Though I never found a complete answer to my question, a good deal of what I learned seemed relevant to the problem of fracture. I filed this basic information away while I waited for the whole puzzle to come together.

Then minor tragedy struck, not once but twice. Near the end of the ski season, my new skis, a Christmas present from my parents, broke. They did not break from some terrible collision or a high-G-force turn; they just fell apart through normal use. I discovered later that this was not a unique experience. The manufacturer had already discovered serious problems with this particular model and agreed to replace them with an improved design. It was a fair offer, but for the time being, I was ski-less. My season was over. Luckily it was time to kayak, and I *did* have a beautiful new boat.

At the time, most kayaks were made from fiberglass and, like mine, many of them were homemade. Homemade kayaks were not necessarily inferior to commercially constructed boats. In fact, homemade kayaks often pushed the technology envelope, being the first to explore the use of new materials. Over the winter our kayak club contacted DuPont and obtained several bolts of their amazing new fabric, Kevlar. We used the Kevlar as a replacement for some of the fiberglass in our boats. The result was extraordinary. Kayaks made with Kevlar were lighter and much stronger than plain fiberglass boats. There was a problem, however, that became evident only when the boats were used. A sharp blow, such as might result from impact with a rock, would cause the Kevlar to delaminate and come apart, like the rind peeled from an orange. After some particularly rough use, my new boat was rendered almost useless by delamination.

First my skis and now my kayak, and to make matters worse, I was unemployed and could afford to replace neither. While looking at my failed kayak, however, it occurred to me that the delamination was fully predictable from the chemistry of the Kevlar fiber. More important, there was probably a way to eliminate the problem. It was incomprehensible that the chemists and engineers at DuPont had not anticipated this. So I called them. As we talked about delamination, I got the distinct impression that these guys knew next to nothing about the relationships between chemical bonds and fracture. I was hit with an epiphany: there were no chapters on fracture in my chemistry or quantum mechanics textbooks, because no one thought of fracture as a problem of chemistry. Surely I wasn't the first.

The more I thought about it, the more I was convinced that there was real value in attempting to deal with fracture using chemistry. I discussed the idea with my adviser, who suggested that I pursue my beliefs in graduate school. He also suggested I consider a course of study other than chemistry. After all, if chemists were uninterested in fracture, it would be unlikely that they would sponsor research attempting to explain why things broke. Instead, he pointed to the disciplines devoted to the study of fracture: metallurgy, materials science, and mechanical engineering. Of these, the name "materials science" appealed to me the most, and I began applying to graduate schools.

ANCIENT ART, ANCIENT CRAFT

CHAPTER

2

There is a ritual that scientists perform when they initially get together. It has a good deal in common with the behaviors of young monkeys or wolves as they attempt to establish dominance. In this wholly meaningless ritual, scientists engage in apparently innocent conversation as they seek to find their place, and everyone else's, in a sort of scientific pecking order. For reasons we can only speculate about, physicists are on the top. Just below are chemists, followed by biologists; then, tightly bunched together, come metallurgists, archaeologists, and paleontologists. At the bottom, one finds sociologists and psychologists. Within each field is a similar hierarchy. Theorists are considered superior to experimentalists, maybe even sufficiently superior to be bumped up a whole notch. Under some circumstances, a theoretical chemist may be judged as equal to an experimental physicist.

When discussing the very real existence of this ladder of scientific dominance, the question always arises as to where mathematicians should be placed. Because no one seems to know what mathematicians do, they are not placed in the dominance hierarchy. Mathematicians are considered separately from ordinary scientists.

As noted, the ritual involves polite conversation, typically beginning with something like "And what is your research interest?" This question is designed to separate research scientists from applied scientists, including engineers and educators. If you are not involved in research, you are not a competitor for research dollars and your place in the pecking order is immaterial. The ritual ends and subsequent conversation sounds much like that between any two people who have just met. "Where do you work?" "How many kids do you have?" If the respondent answers with some research interest, the ritual continues.

When asked the question, I usually say, "I am a quantum chemist." This answer places me in the category of "theoretical chemist"—and my position in the pecking order is established. When I want to have fun, however, I say, "My research is concerned with the study of why things break." Usually a look of satisfaction appears in the questioner's eyes as he says, "Oh, so you are a mechanical engineer (metallurgist, ceramist, or materials scientist)."

Now the fun begins as I say, "No, I study *why* things break, not *when*."

The questioner is now doubtful. I sometimes break down at this point and explain in detail my research; but I have been known to milk the when/why distinction until my questioner is just fed up and moves on to someone who obeys proper etiquette when engaged in the pecking-order ritual.

It is a common misunderstanding, confusing *when* with *why*. What people really understand is *when* things break. Fracture, as with almost every other phenomenon, is composed of two parts, cause and effect. The question of *when* deals with the cause, while that of *why* deals with the effect. An engineer can control the phenomenon with an answer to either question. For example, knowing that a drinking glass will break when it is dropped, an engineer could carpet the floor and avoid a cause of fracture. Similarly, the engineer can change *why* the glass breaks by changing the material from which it is made. A tin cup will not break when dropped. What intrigues me is that so much of mankind's development in history, art, science, and economy is intimately intertwined with the fact that some things break while others bend. Despite the fact that throughout history we have worked to develop a nearly complete answer to the question of *when* things break and bend, we have only the most rudimentary understanding of *why* they do so.

Our earliest ancestors hunted for stones that by happenstance had broken to produce a sharp edge. These sharp-edged stones were invaluable for removing meat from bone and then cracking those bones to remove the rich and nutritious marrow. For how many millennia those hominids relied on found stones is not known, but about 2.8 million years ago some member of the hominid species *Homo habilis* smashed one rock against another, breaking the rocks to produce a cutting edge. The first technology had been discovered, the technology of fracture.

The practitioners of this technology were the first engineers, rock-breakers. They possessed certain basic knowledge concerning when and how things broke. They knew, for instance, that not all stones would break to produce sharp edges. In fact, useful stones were comparatively rare: flint, obsidian, and petrified wood were among the few that fractured concoidally (in shallow concave fractures) to produce a sharp cutting edge. The rock-breakers had learned the most basic law of technological development: the way a material fails can define its use. To be effective, these first engineers had to learn to discriminate among various types of stone, and by 2.6 million years ago, in what is now Ethiopia, they had clearly succeeded. Found here were the first primitive stone tools, simple flint cobbles made by fracture to produce a single sharp edge used for cutting and scraping.

At nearly the same time, ancient hominids discovered they had the ability to make fire. Undoubtedly this discovery, too, was made by the rock-breakers. Sparks are produced when flint strikes against pyrite. The sparks will ignite dried moss and fungi, which can be used to ignite larger pieces of kindling. With the discovery that striking rocks together could be used to make fire, fire was turned to the use of the rock-breakers. Before large pieces of flint could be worked into tools, they had to be broken into manageable sizes. This was accomplished by heating the stones in fire and then cooling them with water or air. The thermal stresses would break the rocks into smaller pieces; in turn, these could be worked, through fracture, into tools.

Over thousands of years, new humanoid species evolved and spread across Africa, as did the art and science of rock-breaking. *Homo erectus,* a heavyset descendant of *habilis,* produced easily recognizable stone tools used for cutting vegetables and butchering meat. Those tools, dating back 2 million years, have been found in Kenya and Tanzania. Among them are double-edged hand axes, which probably made it possible for our ancestors to hunt as well as scavenge larger animals. As suggested by one particular find in Kenya, by 700,000 years ago the rock-breakers had even developed an approach to mass production. Hand axes of the same length but different breadths were discovered. These are believed to have been made using some sort of template.

By 90,000 years ago, fully modern humans had left Africa and moved into the Middle East and Europe, where the rock-breakers further perfected their technology. They could extract several different tools from a single piece of flint; no longer did the shape of the stone dictate the shape of the tool. The craftsmanship of these ancient engineers was first discovered in a nineteenth-century excavation in the Paris suburb of Levallois, from which this particular style of rock-breaking takes it name. Modern attempts to reconstruct the most complex Levallois tools require as many as 111 blows to form the tool, followed by a precise blow to separate the tool from the parent stone. The process of trial and error that led to the first Levalloisian tools must have been incredible. The ability to pass on the information underscores the level of intellect involved.

Making these tools required a detailed knowledge of the fracture properties of flint, and acquiring this knowledge would necessitate a period of training. More significantly, teaching the skills necessary required a language. Whether the knowledge was imparted through hand signals or spoken language, the craft of rock-breaking required millions of years of research and development, and was not learned anew each generation; in each of these master craftsmen resided eons of experience that was either passed on or lost forever. The only guarantee that the experience could be passed on was the development of language. Modern experts have estimated that a minimum of 250 signs were required to describe the Levallois tool-making technique. The first schools may have included intensive study of the way flint breaks, and the first language might have been filled with words describing how and when this happens. Imagine the mothers of early modern humans rousting the kids out of bed and off to rock-breaking school.

The tools made possible by rock-breaking technology allowed early humans to spread across the globe. Accompanying this expansion of range came commerce, and because good flint is not common, it became one of the first trading currencies. Its intrinsic value was derived only from the way it *breaks.*

Maximizing this relatively scarce resource was the driving force propelling the technology of rock-breaking. Using the "punched blade" technique, arising about 50,000 years ago, a skillful craftsman could produce twenty-five times more edge from the same size stone as using the Levallois technique. In the punched-blade technique, a single piece of flint was shaped into a smooth cylindrical core and cut flat along the top. A blow to the flat of the core would produce a sliver, with subsequent blows removing more slivers, much as coins are removed sequentially from a roll. These nearly circular blanks could be further worked to produce 130 different tools, from adzes to projectile points, needles to scrapers. The technology of the rock-breakers had reached its zenith.

Relatively recently, a profound transformation occurred in the mind of the rock-breaking technologist, a transformation that truly altered human development. For nearly two million years the rock-breakers had made utilitarian objects. They had learned to couple their hands with these objects, developing the skills to make ever more complex and sophisticated tools. At some point they began to take pleasure in their skills, and to exercise and perfect their technique simply for the pleasure it gave them. Deriving pleasure from doing things that humans can do is what truly distinguishes us from other animals, not simply the use of tools. Chimpanzees make tools to extract termites from logs, but there is no indication that they take pleasure in the manufacture of these tools. In contrast, the rock-breaking craftsmen of 50,000 years ago made tools with an edge sharper than need be. These tools could be used for purposes for which they had not been designed. In turn, the craftsmen derived pleasure from perfecting these tools and the cycle of technological advance was created.

It is the pleasure we derive from our skills that is the basis for all art and science; and, as artists and engineers, the rock-breakers played a pivotal role. The first objects produced by the rock-breakers that had no apparent utilitarian value were carved stone figurines known as "Venuses." Venuses first appeared in the archaeological record about 20,000 years ago in what is now southern Europe. They are small rounded and smoothed stones clearly depicting the feminine form. The skill to smooth stone into desired shapes provides a new way to look at the material. Where for years the emphasis had been on the reduction of a stone to uncover the tool within, now stones could be assembled to create new forms. Stones were cut, smoothed, and assembled into walls, the walls into houses, the houses into cities, and the cities into civilizations. The rock-breakers were turned to architects and stonemasons. Their art and craft are preserved in the skeletal remains of the ancient cities and monuments of the Romans, Greeks, and Egyptians.

After two million years of observation, learning which stones would break concoidally, where they could be found, how to maximize the number of cutting edges obtained from each scarce stone, and how to shape and smooth them, the stone-breaking artist/engineers faced serious problems. The tools, requiring so much effort to produce, had limited lifetimes. The scrapers and projectile points made from the valuable stones could break after only a few uses. The same property that made it possible to shape these stones into useful objects, concoidal fracture, also limited their lifetimes. A sharp blow to a carefully crafted projectile point (a not-unexpected occurrence for something designed to be thrown or thrust at animals concealing hard bones), and it was broken, useless. The demands of users had exceeded the available technology.

Surely, sometime in the last 50,000 years a rock-breaking engineer fashioned a truly exceptional stone tool, bringing two million years of experience to bear. Then, in its first use, the tool broke and the craftsman wondered, just as I did when looking at my marbles, "Why?" With an answer to this question, the toolmaker may have thought, fracture could be turned on and off: on when the tool was being formed, and off when it was to be used. This craftsman knew everything there was to know about when rocks broke, but was unable to answer the question of why they did so. Without an answer, another solution had to be found to the problem of fracture, a boon for fabrication but a curse when it came to use.

The dilemma undoubtedly triggered the next big leap in human technical evolution. About 10,000 years ago, metals were discovered and used. These materials did not break, at least not in the manner of stones; they bent. The transition to metals was not a fast one. The original metalworkers relied on metals found in their pure form for their raw materials, and such finds are far less common even than flint. The most commonly found metal is native copper. More rare is meteoric iron. For thousands of years these metals were hammered and worked to produce small tools.

Then someone discovered how to separate copper from its ores, making it possible to mass-produce the metal.

Though only a conjecture, it's widely believed that potters first discovered that red copper metal could be extracted from the green stone malachite. Though stones had their applications, they were of limited use as cooking pots. The clay from the riverbank, however, could be readily fashioned into a container, and when dried in the sun, it would hold its shape as well as whatever was put into it. At some point a sun-dried clay pot was used to cook something. Surprisingly, the pot came from the fire harder and more impervious than it had gone in. From this observation grew the potter's craft.

Within a few thousand years the potter would recognize that hotter fires produced harder and stronger pots. To get a hotter fire, bellows were needed. Using a bellows, and the right choice of wood, very hot fires were easily achieved. The craftsman, however, would not be satisfied with only the production of more-useful pots. Just as the rock-breaker took pleasure from the doing, so also would the potter. Taking pride in the work, he or she would wish to adorn it. Grinding malachite and mixing it with water to produce slurry, the potter created a green glaze. When fired, however, the pot did not come out with the green design intended by the potter. Rather, specks of reddish metal adorned the new pot. There was no question what this metal was; it could easily be recognized as the scarce commodity so useful in the manufacture of long-lasting tools. Thus began the large-scale extraction, or smelting, of copper from its ores, beginning about 7,000 years ago in Persia and Afghanistan.

Suddenly humankind had access to a material with none of the disadvantages of stone. It could be drawn, hammered, and cast; used to make tools, ornaments, or vessels; and recycled to make a new object just by throwing it back in the fire. It had one disadvantage, however: copper is a soft metal and will not take an edge. The amazingly sharp edges produced by the rock-breakers could not be achieved by the coppersmiths. Copper did not make good cutting tools. Again, the coppersmiths had learned, just as had the rock-breakers before them, the way a material fails limits it uses. Humankind was in need of a material that could be shaped as a metal, and yet could be made to take an edge like flint.

By the fourth millennium B.C., some coppersmiths were beginning to manipulate the purity of their copper to change its properties. The metallurgical arts were emerging. Like the craft of rock-breaking, early metallurgy was discovered by accident. Copper ores contain trace amounts of other elements such as silver, iron, lead, arsenic, tin, and antimony, which are difficult to remove during smelting. The relative proportions of these trace elements often vary from one ore source to another. The copper smelter learned by observation to associate different properties with the ore from which the copper was smelted. Through some unknown thought process, the smelter discovered not only which ores gave rise to which properties, but also how to increase the concentration of the element responsible for these properties. The first element found in smelted copper, in significantly larger proportions than is typically found in natural ores, was arsenic. Though arsenical copper behaves similarly to pure copper when smelting or casting, it behaves quite differently when worked by hammering, where it becomes much harder.

As the coppersmith was learning to make metals harder and more durable, he also discovered that some ores yielded metals that melted at lower temperatures and flowed better in a molten state than did pure copper. The element responsible for this remarkable improvement in the properties of copper was tin; and as with arsenic, tin makes copper harder. The ideal proportion of tin was found to be about ten percent, producing an alloy known as bronze. The first true bronze appears in the archaeological record about 3000 B.C. in Mesopotamia. Within a few hundred years, however, bronze objects began to appear across the Near East, from Egypt to Cyprus.

Because of the lower melting point, improved flow, and increased hardness of bronze, items made from it were easier to fabricate and able to hold an edge longer. A material intermediate to stone and copper had been created, and it rapidly supplanted copper as the metal of choice for tools, weapons, utensils, and works of art. Copper and tin became the strategic materials of the ancient world. The civilizations that controlled these metals controlled the way things broke, and thereby dominated the ancient world.

Great deposits of copper ore were found on the island of Cyprus. So vast were these deposits that only recently have the copper mines in the center of the island, at Skouriotissa, closed down. Beginning about 2000 B.C., these mines supplied copper for the bronze-reliant technologies of the eastern Mediterranean, and continued to provide copper to the Roman Empire. In fact, the very name of Cyprus is derived from the Roman term for copper, *aes cyprium*. During the Bronze Age, the copper from Cyprus was distinguished by its peculiar shape. Ingots of nearly pure copper, each weighing up to seventy-five pounds, were shaped into the form of "oxhides" with armlike projections on either side. These were shipped across the Near East. The shape may have made the ingot easier to carry, or it may have been a tribute to the bull, the sacred animal of the Cypriots. Either way, the archaeological evidence indicates the extent and importance of copper smelting in Bronze Age Cypriot society. Exploration of the ancient city of Kition has uncovered a complex of temples connected by doorways to copper-smelting workshops. Extensive stores of fine copper and bronze artifacts have been found among the buildings. The proximity of the workshops to the temples underscores the high priority placed on the industry.

While sources of copper are distributed across the Mediterranean, the source of tin supporting the bronze industries of the ancient Near East is unclear. There are no known sources of tin near any of the ancient bronze-making centers. Cyprus was devoid of tin, though there is some evidence that it was a distribution hub for tin arriving from distant sources. Most authorities agree that at the height of the Bronze Age in the Near East, tin was imported there across a network stretching from Spain and perhaps even Cornwall to the west, and from at least as far as India and possibly even farther east. Whatever the source of the tin supporting the Bronze Age technologies of the Near East, the peoples of the region were dependent on its continued supply.

It was also during the Bronze Age that art became truly independent from toolmaking. Until this time the skills of the toolmaker were turned to art. That changed however, as the skills of the artisan developed independently from the need to make utilitarian objects. The low melting point and exceptional flow properties of molten bronze allowed it to be cast into objects of enduring beauty. Nowhere is this seen more clearly than in the Far East, where magnificent cast-bronze artifacts dating back to the middle of the second millennium B.C. have been found. These ranged from small yet incredibly graceful vessels to huge cast cauldrons weighing almost a ton. These massive cauldrons are adorned with rich and intricate relief patterns cast directly into them.

In the Near East, metalsmiths had perfected other techniques for working metals including silver, gold, tin, and lead, but by far it was copper and its alloys, mostly bronze, that dominated metal production. In this part of the world, metals were worked by forming with hammers, hearkening back to the skills developed by the rock-breakers. With these techniques the metalsmith formed intricate tools and works of art. Fabulous examples of the metal craftsmanship of the second millennium B.C. are found in the objects uncovered in the tomb of the Egyptian king Tutankhamun. Three gold coffins held the young king's body. Each was made by assembling solid plates of hammered metal. The burial mask, made from beaten gold inlaid with blue lapis lazuli, shows not one of the thousands of hammer blows that went into its forming. Among the gold and jewels of the tomb was a singular item: found beside the king's body was a dagger with a gold hilt and a blade of virtually untarnished iron. How this ancient blade was protected against corrosion is not known, for its composition of trace elements has never been determined.

Iron has been available in its meteoric form since the beginning of the copper age. However, at the time of Tutankhamun's death it was hardly used, for it was considered an inferior metal. Iron could not be melted in Bronze Age furnaces, and so could not be separated from its ores as molten iron. It could be partially separated by heating the ores to form an iron "bloom" containing a high concentration of iron mixed with slag (oxides of silicon and other impurities). This bloom could then be repeatedly heated and hammered, driving out the slag to produce an almost pure form of iron known as wrought iron. In this form, however, iron is soft, and though it can be hardened by hammering, it still will not hold an edge as well as bronze does. In addition, unlike bronze, which forms an oxide that inhibits further oxidation (this is called a passive oxide layer and is responsible for the beautiful patina on bronze artifacts), iron continues to oxidize, and rusts away when exposed to moisture. Bronze weapons made 4,000 years ago, when cleaned, are as shiny as the day they were forged. In contrast, iron weapons from the fourth century B.C. have corroded almost to the point where they cannot be recognized. Despite all of these disadvantages, iron would replace bronze by the beginning of the first millennium B.C.

This unlikely transition began around 1200 B.C., when all of the eastern Mediterranean was thrown into turmoil. The cosmopolitan community, in which people traveled freely, was invaded by fierce European tribes that swept down from the Balkans into Greece and across Crete and Cyprus. The trade routes, which had been fairly stable for nearly two thousand years, were interrupted and the tin supplies to the bronze-making cultures of the Near East were cut off. The Bronze Age ground to a halt in only two hundred years. If the technological advances that bronze brought were to continue, a new material, which was neither too soft and bendable nor too hard and brittle, had to be found. The only candidate was iron.

Prior to the total collapse of the Bronze Age, there is some evidence that the Hittites had discovered the secret of producing "good iron." Ancient clay tablets indicate that their success in war and the resulting expansion into the Levant and Mesopotamia between 1400 and 1200 B.C. was aided by superior weapons, made from iron, harder than hammered bronze, that broke the bronze shields of their enemies. In a pattern that would be repeated, even down to modern times, the knowledge of how to make things harder, stronger, and more resistant to fracture was guarded from the rest of the world because of the military and economic advantages it offered.

Just as today, where such knowledge is held by the community of scientists doing weapons research, at the time of the Hittites the knowledge was probably held by a subjugated tribe of skilled metalworkers called the Chalybes. They lived along the shores of the Black Sea, in what is now Armenia. With the invasion of the Near East by the European tribes, the Hittite civilization collapsed. At the same time the secrets of the skilled Chalybian metalsmiths may have been revealed. Regardless of the cause, after 1200 B.C. the use of iron, which had once been restricted to the Hittites, spread quickly across the eastern Mediterranean.

The secret of making good iron was probably discovered accidentally. When making wrought iron, the iron bloom is repeatedly heated and then hammered. The temperature of the bloom iron cannot be allowed to drop below about 800° C. or it becomes too difficult to work. The bloom must be continually placed in a charcoal fire where it is exposed to carbon from carbon monoxide gas. Some of this carbon will diffuse into the iron surface, creating an alloy much harder than pure iron. This alloy is in fact steel, and the process of alloying the surface of iron with carbon is called "steeling."

While pure iron is softer than bronze, the effect of steeling is dramatic. Even 0.3 percent carbon makes the iron as hard as or harder than bronze. At 1.2 percent carbon, the steeled iron is much harder than bronze and, if then cold-hammered, develops twice the strength of cold-hammered bronze. By steeling just the edge of a sword, it could be made very sharp and able to hold this edge through repeated use. Unlike bronze, the shaft would remain soft and forgiving, absorbing the blow of an adversary instead of breaking. Not to restrict attention to weapons, all sorts of tools were made more efficient by steeling, from picks and plows to saws and knives. For the first time humankind had a material that, when needed, could be made nearly as hard and breakable as stone or as soft and bendable as copper.

Using carbon to influence *when* iron breaks constituted the foundation of technology for nearly three thousand years. Iron was hammered, rolled, cooled, heated and reheated, and then hammered some more, all in an attempt to exert more control over the fracture properties of iron and steel. Though it was not understood, all of this hammering, heating, and cooling did little more than change the distribution of the carbon atoms among those of the iron, which changed when iron broke. Finally the science of metallurgy would emerge and put everything on a solid footing, explaining that a particular alloy of carbon and iron breaks when the stresses reach a critical level. But the absence of a scientific understanding does not appear to have had much effect on the development of sophisticated steelmaking technology.

In India, a technique for making steel with a uniform distribution of carbon produced a superior form of the alloy known as *wootz*. Highly prized, this steel was traded around the Mediterranean for centuries. The short swords of the Romans were made from it, as were the scimitars of the Moors in the sixth century A.D. Wootz steel was made by sealing wrought-iron and plant material, which served as a source of carbon, in clay crucibles and then heating it in a pit filled with charcoal and fitted with bellows. In this way the iron would melt and the carbon from the plant materials would become uniformly distributed in what had become molten steel. In turn, this was poured into stone molds to produce characteristic wootz ingots.

Steelmaking in the Far East is exemplified by the sword and armor of the Japanese samurai. The making of these implements has been going on in one way or another since about 800 A.D. Unlike wootz steel, the distribution of carbon in the samurai sword results from a layered structure. A billet of wrought iron is forged, then cut in half, doubled over, and forged again. When this is repeated fifteen times, it results in 32,000 layers. The final forging is shaped into the sword and then covered with terra-cotta of varying thickness, thin at the edge, thicker at the shaft. The clay-clad sword is then heated until it "glows the color of the morning sun," at which point it is quenched in water, where, owing to the clay's varying thickness, the edge cools more rapidly than the shaft. This layering and subsequent heat treatment makes a sword with an incredibly hard edge but a forgiving shaft.

The ability to control the arrangement of carbon atoms in iron, whether understood or not, opened up an almost unlimited potential for technological development. Iron was coupled with the power of steam to make steam engines. These engines drove hardened steel tools to forge and then cut steel into the components of railroads, steamships, drilling rigs, skyscrapers, and bridges. Through trial-and-error investigation, the distribution of carbon in each component was chosen so that the *when* of its failure was most appropriate for its function; the breaking point of a steam boiler must be different from that of a railroad track. Taking what was learned about the control of fracture in iron and steel, and combining this knowledge with a little enlightened empiricism, metallurgists found ways to change the fracture properties of alloys based on metals other than iron. With these new alloys came new technologies.

Yet, even with new alloys and the knowledge gained through empirical investigation, there were problems. Things were breaking unexpectedly. Though sometimes the steel bits used in oil and gas drilling would last for days, even while drilling through the hardest rock, at other times they would break with their first use. The British troops in India found that brass rifle cartridges cracked during the monsoon season, even when carefully protected from rain. Whole ships broke in half for no apparent reason. The problem, simply stated, was that the experience gained from thousands of years of trial and error provided no predictive power.

At the beginning of the twentieth century, emerging technologies were demanding reliable materials. Developing those materials required that the art of controlling the *when* of fracture become a science.

ANCIENT SCIENCE

CHAPTER

3

April 28, 1988, Aloha Airlines Flight 243 prepared for takeoff, just as it had many times before. The doors were sealed and the flight attendants instructed the passengers, as they had hundreds of times in the past, what should be done in the event of an emergency. "If the cabin should depressurize, an oxygen mask will drop. Place the oxygen mask over your mouth and nose, and breathe normally until told to remove the mask by a flight attendant." Most of the passengers paid little attention to the familiar message as they anticipated a short, uneventful flight. The Boeing 737 rolled down the runway, gained speed, and lifted off.

As the plane climbed from the Hilo airport, air pressure outside the plane and in the cabin began to drop. This caused the passengers' eardrums to expand. They responded to the annoying sensation with a yawn and a tightening of their neck muscles, allowing the pressure on their eardrums to equalize to that of the cabin. At about 5,000 feet, where the air pressure is about 80 percent that of sea level, the cabin pressure was held steady while the plane continued its climb and the external pressure continued to fall. The pressure differential between the inside and the outside of the cabin forced the atoms of the aircraft's aluminum skin to move ever so slightly farther apart. Near the front of the aircraft, the atoms at the tip of a crack began, once again, a cycle of expansion and contraction started months—even years—earlier. To the passengers and crew, the flight appeared to be progressing normally.

On this short island-hop to Honolulu, meals would not be served, and the flight attendants moved about the cabin routinely dispensing drinks and snacks. A few passengers had fallen asleep, lulled by the drone of the jet engines and the somnolent effect of the slightly reduced air pressure. But this same change was producing a very different effect on the atoms at the tip of the crack, prodding them gently awake.

Through hundreds of pressurization and depressurization cycles, this crack had opened, closed, and opened again. Like the growth rings of a tree, its history was clearly written on its surface. If viewed with a low-power optical microscope, each cycle would appear as a single steplike pattern on the surface of the crack, with the riser of each about the same height, but the treads getting longer with each cycle, the atoms at its tip moving farther apart to compensate for the change in pressure.

Today would be different. As the plane reached its cruising altitude of 24,000 feet, the pilot pitched the aircraft from climb to level flight. This maneuver stressed the fuselage, and the atoms at the crack tip separated beyond the point of no return; the crack reached its critical length and began to run. Driven by the pressure differential, cabin air rushed through the crack, threatening to further enlarge the opening and cause the cabin to depressurize explosively. The Safe Decompression design of the Boeing 737 kicked in as a designed weakness, a decompression flap, tore loose to allow the controlled release of internal air, which rushed through the flap at over 700 mph.

Inside the cabin, the passengers and crew were still unaware of the unfolding events. Less than a second had elapsed since the crack had gone critical. As a flight attendant reached to retrieve a cup from a passenger, the decompression flap opened above her. Unrestrained, she was sucked into but not through the ten-inch-by-ten-inch opening. The flight attendant's impact produced tremendous stresses and the fuselage began to crack like an eggshell. Eighteen feet of the forward cabin, along with the hapless flight attendant, were torn from the aircraft. Luckily the pilot was able to retain control and land what had become a flying convertible.

Fortunately, such occurrences are not common. Rarely

do cracks in important components, such as the fuselage of an aircraft, reach critical lengths, the point at which cracks become unstable and simply run. As part of the material development process, new alloys are studied to determine, for a given stress level, the size of the critical crack. Knowing this length, we can design things that are reliable, and seldom are there catastrophic consequences associated with premature failure. At any rate, this is the state of affairs now, but for the previous seventy-five years, unexpected failure was the rule. There were no life expectancies for manufactured products; they were simply used until they became useless. In some cases they would fail catastrophically, causing loss of life or serious injury.

The story of the transformation, which took unexpected failure from the rule to the exception, began 2,500 years ago with the Greek philosophers. At that time the big mystery in natural philosophy—what we now call science—revolved around the coexistence of two everyday phenomena we take for granted, conservation and change. We know that the stuff of the universe can both be conserved and change. The water in a puddle doesn't disappear on a hot, dry day; it is conserved and changes into vapor through a process called evaporation. The Greeks didn't know this, and there was really no reason they should. Everyday experience is filled with stuff apparently coming into existence and then vanishing. Rain falls from the sky, forming lakes and rivers. In the dry season these can vanish, leaving no trace. The dramatic and manifold changes of the world suggested to some that things could spontaneously arise from nothing. Yet it also appeared that the total stuff of the world was constant.

Copper ore was refined to copper metal in a process that left behind slag. The total amount of slag and refined copper weighed roughly as much as did the original ore. The copper, once formed, could be shaped into different objects, but they would all weigh the same. How could there be both change and conservation? So puzzled were many of the natural philosophers of the time, they speculated that change itself was an illusion, occurring only in the mind.

In what is probably the single most important theory based on pure thought, Empedocles proposed a solution to the problem. In about 445 B.C. he speculated that the stuff of the world was made of different "roots." The roots were seen as being elemental, in that they were original and had always existed, as well as being indivisible or atomic (*atomon* is the Greek word for "indivisible"). Change was caused by roots mixing and coming apart under the influence of the two opposing forces of nature, "love" (attraction) and "strife" (repulsion). Different substances came into being, as the different types of roots combined in different proportions.

What an amazing insight! In one concise statement, Empedocles proposed an atomic theory of matter, which embodies the essential features of the modern theory of the same name. Both theories account for the huge number of observable substances in the world, with each composed of atoms of different elements in specific proportions. As an explanation of the coexistence of change and conservation, this theory was dead on.

Some modern historians have minimized the scientific achievements of the Greek atomists. Bertrand Russell, in his

History of Western Philosophy, put it this way: "By good luck, the atomists hit on a hypothesis for which, more than two thousand years later, some evidence was found, but their belief, in their day, was nonetheless destitute of any solid foundation."

I question Russell's assertion that this milestone in scientific thought was achieved through sheer luck. Rather, it appears clear that the Greeks formulated a theory consistent with the world they observed. And in this world, the rich varieties of man-made things were constructed from a finite number of raw materials, not the least of which was stone. From cut and shaped stones, all manner of things could be built. It is not surprising that a theory developed that imposed on the microscopic world the same structure as that observed in the macroscopic world. What would be more sensible than to assume that all of creation was assembled from some set of primary building blocks? And though our powers of observation have expanded far beyond those of the ancient natural philosophers, theories are still based on observation.

Unfortunately for scientific progress, Empedocles' theory of change was also used to account for the properties of matter, which were seen to be a consequence of the combined properties of the roots from which the substance was composed. Rather than a correct theory of change, the atomic theories of the Greeks were used to explain constitution and properties. Consequently, two thousand years would pass before serious questions regarding the nature of atoms would once again arise.

Empedocles held that there were four kinds of atoms:

earth, air, fire, and water. Things of the world, though not necessarily of the heavens, were made through the combination of these atoms in various proportions. However, at the time, the words *earth, air,* and *water* had much more general meanings than today. Earth was a term applied to a wide variety of solids, air to gases, and water to liquids or things that could be fused, including metals. As a theory of change, Empedocles' choice of the four elements was clearly not arbitrary. The elements earth, water, and air represent the three states of matter: solid, liquid, and gas. The transformations between the states of matter are among the most conspicuous and everyday examples of change, often the result of heating by fire. Fire, earth, water, and air—as elements—put change at the center of Empedocles' atomic philosophy.

There is only fragmentary evidence to suggest the extent to which Empedocles sought to apply his theory of change and describe the composition of various compounds. It is known that he believed bone was made of atoms of fire, water, and earth in the proportions 4:2:2 respectively. Over the centuries, though, the emphasis clearly shifted from the problems that change presented toward a theory of composition. At the same time, the general picture of the four elements, as atoms of change, was lost and the elements were associated with specific substances. It was the great French scientist Lavoisier who, in the middle of the eighteenth century, recognized that water and air were not indivisible, and that fire was a process, not a substance. These realizations marked the birth of the modern science of chemistry—but that is another story altogether.

Notable contributions to the atomic theory came from Democritus, who some consider the father of the atomic theory because of the influence his writings had on subsequent philosophers. Democritus pictured atoms as being in incessant motion in a void. In this respect, he proposed a model very close to the modern kinetic theory of gases. Aristotle's chief concern with atomic theory was to explain the properties of a compound from those of the atoms constituting it. His atomic theory survived almost intact into the eighteenth century A.D. He argues in his treatise *On Coming-to-Be and Passing-Away* that every tangible quality (property) of a body can be represented as a position on a scale between two opposites—for example, hard and soft, or rough and smooth. Some of these opposites were considered dependent, however, as they could be reduced to others; hard and soft, for example, could be treated as modifications of wet and dry. Aristotle considered the minimum set of opposites needed to account for all tangible qualities to be two, cold/hot and wet/dry. By design or by chance, four elements were needed to provide all possible combinations of these two pairs of opposites. So, as with the earlier theories, all substances were seen as composed of the four simple elements. In turn, the properties of each of these resulted from a combination of two of the four primary opposites. Earth was cold and dry, water cold and wet, air hot and wet, fire hot and dry. This theory neatly explained some of the more common examples of change—say the transformation of water on boiling to steam. In the Greek view, this would appear as water changing to air, that is, the cold and wet changing to the hot and wet. The agent of this change was fire transforming the cold to the hot.

To a modern reader, Aristotle's view that materials' properties arise from polar opposites seems naïve, but in fact he was very close to hitting on the true origin of properties. Aristotle, of course, knew nothing of electricity and magnetism, but if he had, it is likely that one of his pairs of opposites would have been the positive and the negative, the proton and the electron. The properties of atoms (in the modern sense of the word) come from the different numbers of protons and electrons they possess, and the properties of molecules arise from the different arrangements of these atoms.

In his *Meteorologica*, Aristotle considered the origins of materials' properties. He chose to describe in detail eighteen qualities of a substance. They were "solidifiable, meltable, softenable by heat, softenable by water, flexible, breakable, fragmentable, capable of taking an impression, plastic, squeezable, ductile, malleable, fissile, cuttable, viscous, compressible, combustible, and capable of giving off fumes." He gives examples of substances that have each of these properties, and then rationalizes their origin in terms of the relative content of the four elements.

Aristotle clearly recognized the importance of the competing and complementary properties governing things bending and breaking; fully twelve of the eighteen qualities he described concerned the competition between fracture and deformation. However, at the time neither Aristotle's atomic theory nor any other could provide a rationale for why some things bent while others broke. The reason is simple: these theories did not explain, or incorrectly explained, what it was that held atoms together in the first place.

Greek science had no notion of the fundamental forces of nature. Nothing was known of the forces of gravity and electromagnetism. In the absence of any concept of the unseen forces that might hold atoms together, the early philosophers pictured molecules as held together via hooks, barbs, or other mechanical means. The shape of an atom was seen as endowing it with properties. Theophrastus, in *On the Senses,* attempted to account for tastes, colors, smells, and so on in terms of specific atomic shapes and configurations. For example, an acid taste resulted from sharp, thin, small atoms, while a sweet taste was a consequence of larger, rounded ones. Over time, properties became increasingly intertwined with atomic shape and arrangement. By the seventeenth century A.D., properties derived from a veritable zoo of differently shaped atoms. There were springy wire balls; needles of acid that could insinuate themselves into a metal; loose structures that bent and tight structures that cracked. Among all the speculation about the shapes of atoms was the image of them as spheres stacked in a regular fashion to form crystals.

Most of the inorganic compounds of the world are crystalline. This may seem surprising, since crystals are usually associated with jewels like diamonds and emeralds, but in fact aluminum soda cans and the steel in your car are also crystalline. To be precise, unlike jewels, which are single crystals, steel and aluminum are assemblies of crystals and are referred to as *polycrystals,* with an individual crystal of this assembly called a grain. There are, of course, many substances in the world that are not crystalline—that fine Waterford or Steuben crystal really isn't—it is glass.

Whether a material is crystalline or glassy depends on the arrangement of its constituent atoms. A crystal is to glass as a Christmas-tree farm is to a forest. Trees are the important constituents of both, but there are some significant differences. Imagine that you have selected a particular tree at a Christmas-tree farm and wish to give the farm attendant directions to find it. You might specify your selection as the fifteenth tree in the tenth row. With only the ability to count, the attendant could find your tree. On the other hand, had you found the perfect tree in a forest, how could you direct someone to its location? A tree in the forest does not have an address like one on a farm; you would need to specify the distance and direction of your tree from some initial point, such as it's twenty-five feet due north of the large boulder.

The need to indicate direction is what ultimately distinguishes the forest from the farm. In the forest there are no preferred directions, making it easy to become lost and begin wandering in circles. If you became lost on a Christmas-tree farm, however, you'd simply walk along a row until you reached the edge of the field, then walk around the field until you came back to your starting point.

Long before it was possible to look at the atoms of a material to see if it was crystalline or glassy, the way it broke gave away its true nature. Some common substances, such as calcite and rock salt and other, more valuable ones, such as diamond and sapphire, break to produce smooth faces, in a process called cleavage. These materials may be fashioned into highly faceted polyhedrons that diffract light and seem to sparkle from within. This aesthetic interplay of light with fracture surfaces is what gives jewels their value. Hence was born an economic as well as an artistic incentive to find the number, combination, and arrangement of facets to maximize the "fire" of precious stones.

Quickly it was discovered that arbitrary arrangements of facets could not necessarily be produced, the facets intersected at fixed angles. The angles would vary from one cleavable material to another, but for each substance, only specific angles could be realized. This observation is easily explained if the substance is a crystal of roughly spherical atoms. To understand this observation, take nine pennies and arrange them on a flat surface. Place three pennies in a straight line; then place four in the hollows above and below this line. Finally, use the remaining two pennies to "cap" these rows, forming a four-sided crystal with three coins on a side. The sides of this crystal join to form only angles of 120 and 60 degrees. Adding or removing pennies does not change the angles formed. As long as pennies are placed in the hollows of two neighboring coins, every polygon formed must have sides that intersect at angles of 120 or 60 degrees. Other crystalline arrangements can be made. Instead of placing pennies in hollows, place them at the high points, such that each penny touches only one coin in each of the rows below and above. The penny crystal now looks to be made of squares where the sides can only form angles of 45 or 90 degrees.

In 1678, the English scientist Robert Hooke recognized

the restrictions that crystallinity placed on the angles between facets, and postulated that at least some substances were formed from packing spherical atoms together into crystals. This insight was necessary to understand the nature of the competition between bending and breaking. As a simple example, consider planes of atoms stacked one on the other like a deck of cards. There are two distinctly different things that can be done with this deck: first, it can be divided into two decks, that is, the cards can be "cut," and, second, it can be fanned, or sheared, by sliding the cards across each other while they remain in one stack. The first of these processes is what happens when something breaks: the planes of atoms simply are pulled apart. The second process is what happens when something bends: planes of atoms slide across each other.

Graphite is an example of a substance with mechanical properties like those of a deck of cards. Graphite is made from carbon atoms tightly bound together into two-dimensional sheets; these sheets are then loosely bound together to form a three dimensional crystal. Because the bonding within a sheet is so much stronger than that between sheets, graphite is said to be anisotropic, meaning it has very different properties depending on the direction in which it is "cut." When pulled in a direction that lies in the carbon sheets, graphite is very strong, making it an ideal substance from which to make tennis rackets, golf clubs, and bicycles. At the other extreme, graphite is used as a lubricant because the weakly bound sheets of atoms shear so easily.

Graphite is unusual. Most materials are not characterized by such extremes of bonding; normally the bonds within a plane of atoms are nearly equal in strength to those between the planes. Also, most structural materials, such as aluminum and steel, are polycrystalline, with the preferred directions in the individual grains randomly oriented. Hence, when a component such as the strut of a bridge is loaded, some of the planes of atoms are subjected to sliding shear forces, while in other grains, the planes are subjected to tensile forces pulling the planes apart.

Just because the planes of atoms are subjected to these forces does not mean that they move. Anyone who has pushed a stalled car knows this to be true. You lean into the car and nothing happens. You then set your feet firmly and push again. Eventually the car starts to move. The same is true for the planes of atoms subjected to load in the bridge strut. For the most part, the strut is designed so that the forces of everyday use will never cause the planes of atoms to slide or come apart, but if, for whatever reason, the forces should exceed this design limit, two things can happen. First, the planes of atoms can slide and the strut will bend, or, second, the planes of atoms can come apart and the strut will break.

I am sure you would prefer that neither of these things happened while you were on the bridge, but if you had to experience one of the two scenarios and you were not an extreme thrill seeker, you would probably prefer to have the bridge bend. At least that way it would remain intact, giving you the opportunity to get off. If the strut were to break, however, the bridge would collapse and you would plunge into whatever abyss it crossed. If the strut is to bend before it breaks, planes of atoms must slide across each other before they are pulled apart.

At first blush, it appears that controlling fracture requires the ability to alter the bonds holding planes of atoms together. The only way to do this is to alter a material's chemical composition, which would change the *why* of fracture. Historically, this was not the course followed; rather we learned how to change *when* planes of atoms slid across each other.

By the early twentieth century, bending was known to be a consequence of sliding of atomic planes. However, deformation was also accompanied by a phenomenon called work hardening, which could not be explained by simple sliding.

My first exposure to work hardening came while I was helping my stepfather install a new sink. At that time, the supplies to the sink were made of copper cut from a roll of soft tubing. These were carefully bent to connect the cut-off valve on the wall to the faucet. Bending the copper so that the connections did not leak required some skill, and my first attempt was a little off. No problem, right? The tube could just be bent again until it was the correct shape. Each time I tried to bend the tube, it required more force. After a few tries to get the shape right, the tubing was too hard to be bent by hand. At this point my stepfather took over, using fresh tubing to form the supplies. If bending were just the sliding of atomic planes, it should be possible to bend something repeatedly, just as a deck of playing cards can be fanned again and again. There is something missing in our picture of deformation.

Work hardening occurs because planes of atoms actually *do not* slide across each other all at once, but move in little steps. One way to picture this movement is by analogy to sliding a rug across the floor. For a small rug, say three by five feet, this is no big deal, you just take hold of an edge and the rug slides. For a bigger rug, it is not that simple. When I take hold of the edge of a twelve-by-fourteen-foot rug and try to slide it, it is like trying to push a car uphill—nothing happens. However, starting at one edge, it is a simple matter to put a small wrinkle in the rug and then push it to the other edge. In the process, the rug will slide a bit across the floor. You can demonstrate the effect with a piece of notebook paper placed lengthwise on a table. Hold the right edge of the paper firmly with your right hand; then push the left edge to the right about an inch. There should now be a little roll or wrinkle in the paper. Push this to the right with your left hand, letting up on your right hand and allowing the wrinkle to pass under. The paper should once again be flat on the table, having moved an inch to the right. The paper did not move all at once, but bit by bit as the wrinkle propagated through the sheet.

Wrinkles in a crystal are called dislocations, and it is through their motion that planes of atoms shear past each other. There is a better way to visualize a dislocation than thinking of it as a wrinkle. In fact, there is a convenient prescription for finding dislocations. Return to the Christmas-tree farm. Imagine, while walking along one edge of the field, that you count thirty-five rows of trees. You then proceed to the opposing edge of the field and once again count the number of rows. This time you find there are thirty-six.

This means there must be at least one dislocation in the field. You can find the center of the dislocation by walking along each row. One of these must end somewhere within the field. This is the center of the dislocation. From the air, the dislocation would appear as a partial row of trees. In a crystal, a dislocation appears as a partial plane of atoms, like placing a half-sheet of paper somewhere within a ream. The edge of the half-sheet is the center of the dislocation. This model is inadequate, however, as it is hard to picture the dislocation moving through the ream of paper. Yet in a real crystal, where the sheets are made of planes one atom thick, a shear force would cause the half-sheet of paper to move, exiting on the top or bottom of the ream and leaving behind a microscopic step on the surface.

Unlike the paper example, most of the grains of a polycrystalline material have no free surface; they are totally contained and bounded by other grains. The surface over which two grains make contact is called a grain boundary. Typically, the atomic planes of the crystals on either side of a grain boundary are not aligned. As a result, dislocations moving across one of these grains are frequently unable to move into the adjacent grain, their motion being blocked by the misorientation of the atomic planes at the grain boundary. The dislocations can only be made to continue their transit of the grain if more force is applied (the force necessary to make dislocations move is called "yield stress"), in essence, pushing it through the grain boundary. With the motion of dislocations blocked, bending (yielding) also stops and a polycrystalline material responds by breaking, as planes of atoms can only be pulled apart. Blocking dislocation motion is the secret to work hardening. In the case of the sink supplies, the dislocations in the copper tubing were initially free to move. With each bend, however, dislocations moved until they encountered grain boundaries or other obstacles, where they piled up and bending was inhibited.

The key here is that the tendency to break or bend—that is, to fracture or yield—is sensitively dependent on a dislocation's ability to move across grain boundaries. The more difficult we make it to get through these blockades, the harder and more susceptible to fracture a material becomes. There are several ways to make grain boundaries effective barriers to dislocations. The most obvious of these is just to make more grain boundaries. The smaller the grains, the shorter the distance a dislocation can move before being blocked. Another approach is to make the material out of two different types of crystals, one in which dislocations can move easily and another where their movement is difficult. Grain boundaries between these two different crystals will be effective inhibitors of dislocation motion.

Though the knowledge of the early metallurgists was gained through trial and error, carefully guarded, shrouded in mysticism and ritual, their art, too, was tied to the ability to inhibit dislocation motion. Cold-hammered bronze is hard because dislocations have piled up at grain boundaries and can no longer move easily. Steel can be made harder than bronze because it is made from two different kinds of crystals, iron and iron carbide. Dislocations that move easily through iron crystals cannot move in iron carbide and thus pile up at the iron/iron carbide grain boundaries. The amazing samurai sword derives its remarkable properties from the tremendous number of these boundaries, formed from the repeated folding of the blade upon itself. This, coupled with quenching of the terra-cotta-clad blade, made the edge cool quickly compared to the shaft. Unknown to the swordsmith, this quenching formed many small grains where the steel was cooled rapidly, and larger grains in the more slowly cooled regions of the blade. The result was a blade that could take and hold a sharp edge, where the many small grains and iron/iron carbide boundaries blocked the motion of dislocations and a softer, bendable shaft, where the larger grains allowed dislocations to move easily.

Early in the twentieth century, 10,000 years of empirically derived metallurgical knowledge were reduced to science. The motion of dislocations was recognized as the mechanism of bending, while suppressing this motion led to fracture. The new science of metallurgy owed much to the earlier atomic theories, which ultimately allowed us to recognize that spherical atoms could be packed together into crystals. In turn, the atoms of crystals could move in two different fashions. One of these must lead to fracture and the other to bending. The important ingredient in controlling the competition between these two types of motion was in the arrangements of grains that blocked or encouraged dislocation motion. So important was the study of these arrangements that they were referred to collectively as microstructures, and metallurgy became a science of microstructure. The central discoveries of this science were thermal and mechanical processes that could be used to fashion desired microstructures, where desirability was to be evaluated by its effect on dislocation mobility.

Metallurgy and, later, materials science would grow beyond their exclusive concern with relationships between microstructure and mechanical behavior. However, fracture and deformation are controlled primarily through changes in microstructure. As in the case of Flight 243, fracture usually begins at the tip of a preexisting crack or flaw. When such a flaw is subjected to a load, sliding forces act on the atomic planes inclined to the crack, while tensile forces act on the parallel planes. To avoid fracture, dislocations must be able to move along the planes inclined to the crack. On the other hand, if dislocation motion is blocked, the planes parallel to the crack will come apart and the crack will run. Clearly, altering the microstructure around a crack will have a dramatic effect on a dislocation's motion and therefore on the tendency of a material to fracture or bend. The important microstructural parameter measuring this competition between yielding and fracture at the tip of existing cracks is called the *critical stress intensity factor*. It is measured experimentally and then used to find the length of a stable crack in a given microstructure. With this information, structures can be designed that will not fail unexpectedly as did the Aloha Airlines Boeing 737.

Consider two different tubes with the same dimensions, one made of glass and the other of copper. The glass tube is the stronger of the two, as it can support a greater weight before breaking. However, if small scratches were to be made on both tubes and they were once again tested to determine which supported the greater weight, the glass tube would be found to break long before the copper one, because the critical stress intensity factor of glass is smaller than that of copper. This is why glass can be cut by simply scratching and then gently tapping it, while copper cannot.

One reaction to a failure like that of Flight 243 might be to replace the aluminum skin of the aircraft with a stronger material. In the extreme, this would be like building an aircraft out of glass; though it would be strong, even a small crack or scratch would lead to catastrophe. Cracks are inevitable, so, rather than attempt to eliminate them, designers have chosen to build from materials where a crack approaching its critical length can be detected and actions taken to repair or replace the affected part. The investigation into the incident on Flight 243 by the National Transportation Safety Board revealed that the crack in the fuselage at the time of takeoff on April 28, 1988, was six to eight inches long, clearly visible, and should have been detected.

The designers had done their jobs; the materials of the fuselage had behaved as expected. The accident was attributed to Aloha Airlines' inability to detect an obvious crack. In fact, after the accident, a passenger stated that as she was boarding the airplane, she observed a fuselage crack. Unfortunately, she made no mention of the observation to the airline ground personnel or flight crew.

Microstructural control of fracture is both powerful and efficient, but it deals only with the *when* of fracture. After all, the critical stress intensity factor tells us that a crack will become unstable when stresses at its tip reach some critical value. Why the crack becomes unstable at this particular stress and not some other is unknown. Yet, because there is such a well-established formalism for controlling fracture, in 1979 only a few scientists saw a reason to ask why, let alone expend time and resources in an effort to answer this question. Fortunately, I was lucky enough to stumble upon those scientists and become a part of their search for an answer.

EMBRITTLEMENT AND OTHER COINCIDENCES

CHAPTER

4

My hunt for graduate schools began in the fall of 1978. It was clear that none of the colleges or universities in Colorado had the combination of expertise that would allow me to study fracture and chemical bonds. This posed a serious problem in my personal life, because my girlfriend, Cheryl, was now firmly ensconced at the University of Colorado, studying psychology. Cheryl wasn't thrilled at the prospect of uprooting her life and beginning again at a new university in another state. However, I did have at least one ace up my sleeve.

Cheryl had spent time in Boston and was enamored of the city. If there was any place to which she might relocate, it was Boston. This narrowed my search dramatically; fortunately, there were a couple of fine schools across the Charles River from Boston, in Cambridge. One of these was Harvard University and the other was a little technical school just down the road, the Massachusetts Institute of Technology, MIT. Not surprisingly, MIT was reputed to have the best materials science program in the world, so it was an easy choice and I soon sent my application to their materials science graduate program.

Though MIT was not the only university to which I applied, because of Cheryl it was my first choice. Seeking every possible advantage, we decided to visit Boston and MIT. I hoped to meet some of the materials science faculty and talk with them about chemical bonds and fracture. Upon learning of our plans, the Materials Science Department helped make the arrangements. They put us in a hotel and arranged an interview schedule for me with members of the faculty and a tour of the campus for Cheryl. I was overjoyed and absolutely filled with anticipation as we landed at Logan Airport in the middle of January of 1979.

My first impression of Boston was how cold it seemed. The television news reported the temperature as about 15° F. For a Colorado boy, this should not have been a big deal; in Denver, subzero temperatures are common. But that fifteen degrees in a calm Boston felt colder than twenty below in Denver. Almost everyone thinks they know why it feels so much colder in Boston. Just ask, and nine times out of ten the answer will be "It's the moisture" or "It's the humidity."

Well that, as my grandmother would say, is poppycock, and I need to digress for just a bit to straighten out this bit of meteorological mythology.

At temperatures below 32° F., the difference between water vapor contained in dry air (zero percent humidity) and saturated air (100 percent humidity) is not that large. Cold air just doesn't hold much moisture. What determines how cold you feel is how fast heat is lost from your body, and this in turn depends on the density of the air around you. That is why a person will die of exposure after only a few minutes in near-freezing water, but can survive for hours, even if naked, in air of the same temperature. In water, body heat is lost very fast becasue the water is dense and can hold a lot of heat. The air, being less dense, holds much less heat and so a body does not cool as fast. Now, Denver is a mile above sea level, whereas Boston is located right on the coast. The air in Denver is only 80 percent the density of that in Boston. That's why it feels colder in Boston; heat is not lost as fast in Denver. It has nothing to do with the moisture.

Because fifteen degrees doesn't feel particularly cold in Denver, I had not brought an overcoat to Boston. When we got off the subway at Kendal Station, the MIT stop, I wore only a sport coat. Cheryl, on the other hand, was properly attired for the weather. At that time the area around the subway station was rundown and looked nothing like I had expected. There Cheryl and I were, looking for anything that resembled the campus of a world-famous university, when two Cambridge police officers drove by and pulled over, apparently checking us out. With a bit of trepidation, we asked the officers for directions. They asked why I was looking for MIT, to which I answered something to the effect that I was a prospective graduate student. This amused the officers no end. They broke out laughing, told us to get in the car, and proceeded to take us on a twenty-minute tour of Cambridge and the MIT campus.

It seemed that what tickled the officers was the idea of a prospective graduate student too stupid to wear a coat. We drove along Memorial Drive, with the Charles River on the left and the massive Romanesque architecture of MIT on the right. As we drove, they pointed out student after student wearing huge parkas, goodnaturedly remarking that MIT students had enough sense to wear coats. After the unexpected police escort, the officers wished us success and deposited us at MIT's main entrance. We could not have asked for a better introduction to the Boston area.

Our introduction to MIT was no less enjoyable. A senior graduate student, Mike Wargo, acted as our host. That morning I was to meet separately with three faculty members of the Materials Science Department, while Cheryl had the opportunity to talk with faculty doing research in perceptual psychology. We had lunch together at the Faculty Club, where Mike entertained us with stories of life at MIT and suggested that Cheryl might wish to investigate the psychology program at Wellesley University. In the afternoon I was to meet with three more members of the faculty. These meetings would start me down the road toward understanding why things break.

In preparation for the trip, my hosts had sent brief profiles of the materials science faculty. From these I was to select the individuals I was most interested in meeting. Professor Morris Cohen's profile was incredibly impressive. He was a member of the National Academy of Engineers and the National Academy of Sciences. He had published many papers on something called martensite; I had no idea what that was, but it sounded good. Among his many publications were several dealing with fracture. That was all I needed; Professor Cohen was one of the people I asked to speak with. The other person I was eager to meet was Professor Keith Johnson. Though Professor Johnson was not studying fracture, he was involved in research explaining the properties of materials in terms of their electronic structure, that is, in terms of their bonds.

Professor Cohen was first up, followed by Professor Johnson, and then a meeting with the department head to cap off the afternoon. When I arrived at Professor Cohen's office, an administrative assistant informed me that Professor Cohen did not meet with prospective graduate students, as it would consume too much of his time. Evidently, virtually every visiting prospective graduate student asked to meet with the distinguished professor. However, Dr. Greg Olson, who worked with Professor Cohen, wished to speak with me. I was shown to an office right next door to Professor Cohen's, where Dr. Olson greeted me.

Dr. Olson—Greg, as I now call him—typifies everything wonderful about good scientists. They exude enthusiasm and curiosity. To scientists like Greg, the whole world is a series of puzzles, and their desire to solve them is downright contagious. To many, such scientists seem more like children than responsible adults. Perhaps this is the reason that good university scientists seldom look their age. To this day, I have no idea how old Greg is. In my mind, he simply doesn't get any older.

Greg's zeal for science was immediately apparent as he said, "So you are interested in bonds." This was a little unnerving, as I had no idea how he knew this. None of the faculty with whom I had spoken in the morning appeared to know anything about my interests. So it took a bit of time to recall that in the application to MIT a short paragraph describing interests had been required. Greg had a copy of my paragraph.

This short paragraph summarized an attempt I had made a year earlier to calculate the strength of a diamond, based on characteristics of the carbon-to-carbon bonds holding it together. In retrospect, the approach reflected a good deal of naïveté about fracture, but I took his interest as an invitation to discuss my philosophy for "making things stronger" by manipulating chemical bonds. Greg quickly waved it off, explaining that strength was probably the best understood and most controllable of materials properties. This control, he pointed out, came through the manipulation of microstructure (at that time something I knew very little about). I began to feel extremely self-conscious. I had come to MIT unprepared, sans coat and with little knowledge of the very subject I was determined to study. Then Greg asked if I had thought about the problem of embrittlement.

It turns out that embrittlement is one of the more serious problems confronting the field. It occurs when a structural material is exposed to impurity elements. Through a number of apparently different mechanisms, these elements can transform an otherwise ductile material into a brittle one. What is startling is that even the most minute concentrations can produce this transformation. A few sulfur atoms per million iron atoms will cause ductile steel to shatter. Hydrogen atoms appear to embrittle just about everything: steel, aluminum, titanium, and the silicon of computer chips. Mercury is another indiscriminate embrittler. In *The History of the World* (78 A.D.), Pliny described its ruinous properties: "so penetrant is this liquor, that there is no vessel in the world but it will eat and break through it, piercing and passing on still, consuming and wasting as it goeth."

On the one hand, stopping mercury-caused embrittlement is often not too difficult. Simply avoid exposing sensitive structural materials to mercury. Several years ago, the instructional information indicating what was and what was not allowed on airplanes showed the extremes to which airlines were going to keep all sources of mercury away from aircraft. These signs, written in English, indicated that among other things, firearms, explosives, and corrosives were not allowed in luggage and carry-on items. Underneath each was an international icon suggesting the meaning of the written English. Below firearms was a circle around a picture of a handgun with a line drawn across the circle. Below corrosives was a similar circle with a line drawn through it, but in this case, in the circle was the picture of a thermometer. These signs have become less common as mercury thermometers have been replaced by their digital and alcohol counterparts.

On the other hand, eliminating hydrogen or sulfur embrittlement is more difficult. These elements are ubiquitous. Hydrogen is a primary component of water and the hydrocarbons from which gasoline is produced. At typical combustion temperatures, small amounts of molecular and atomic hydrogen are generated. These rapidly diffuse into structural components, making them brittle. Sulfur is a common impurity in coal, natural gas, crude oil, and iron ore. Through the iron ore or coal, sulfur may be incorporated into newly smelted steel. Even sulfur-free steel may become brittle if used in a sulfur-rich environment, posing a serious problem for the pipelines transporting sulfur-rich crude oil, or for the drill bits and other components used in oil and gas exploration.

Embrittling elements do not work their mischief on microstructure; the microstructure of steel, prior and subsequent to hydrogen exposure, is the same. In 1979, as I was sitting in Greg's office, it was believed that embrittlement was caused by changes to the chemical bonds at the tips of cracks where fracture initiates. In some unknown way, these elements changed the bonds, causing either the suppression of dislocation motion on the planes inclined to the crack, or promoting fracture on the planes parallel to the crack. Either way, the affected material would become brittle. Before it would be possible to eliminate embrittlement in affected materials, we had to understand why and how some elements produced their deleterious effects. As Greg explained all this, my spirits rose.

On that day, Greg began to describe a little of his research on sulfur and phosphorus embrittlement. He explained that sulfur and phosphorus atoms in iron and steel are mobile and move around like checkers on a checkerboard. They hop from one site to another, and continue hopping until they reach a grain boundary—and there they stop. In this way, even if the number of these atoms in the steel is low, the concentrations at grain boundaries can be quite high. In essence, the grain boundaries attract both sulfur and phosphorus. Around Greg's office were models of iron grain boundaries built from what were essentially Tinker Toys. As he described the motion of sulfur atoms in iron crystals, he would grab one model and then another, to indicate where he thought the embrittling atoms might be located, and speculate on what might be happening to the chemical bonds. These were represented as little sticks in the Tinker Toy models. I was swept up by Greg's obvious passion and the mystery he presented. What exactly was it that phosphorus and sulfur did to the grain boundary bonds of iron? Why did this make steel brittle? Like any good mystery, this one needed a solution.

Though the concentrations of undesirable elements in steel or other alloys can be reduced, through various chemical methods during processing, deleterious concentrations will remain. As a result, other methods have been found to lessen their effects. The most common approach is to "getter" the bad atoms by giving them targets that are more attractive than grain boundaries. For example, often manganese is intentionally added to steel. The manganese reacts with sulfur to form a stable compound—manganese-sulfide. The sulfur atoms in this compound are immobilized and cannot move to iron grain boundaries.

The sinking of the *Titanic* is the most famous and conspicuous example of the terrible consequences of sulfur embrittlement. Almost from the day the great ship sank, it was assumed that upon hitting the iceberg, a terrible gash, nearly one hundred meters long, was torn through six of the sixteen watertight compartments of the "unsinkable" ship. Only after the wreck was located, and steel specimens from the hull were recovered, did the true story become clear.

It is likely that all, or most, of the steel used in the hull of the *Titanic* came from open-hearth furnaces in Glasgow, Scotland. These furnaces were acid-lined and, unlike the basic open-hearth furnaces of the time, did not react with sulfur and phosphorus to remove these impurities. Elemental analysis of the recovered hull material confirms this conjecture, showing a sulfur content nearly twice and a phosphorus content nearly four times that of modern steels used in similar applications. Not only was the sulfur content of the *Titanic* steel high, but the manganese content was low, with a manganese-to-sulfur ratio of 6.8 to 1, compared to similar modern steels with a ratio of 15 to 1. These elemental compositions conspired to make the hull of the *Titanic* brittle, but unfortunately not sufficiently brittle to be noticed by the workers constructing the ship at the Harland and Wolff shipyard in Belfast, Ireland.

As the sulfur content of steel is changed, it does not suddenly transform from ductile to brittle. In fact, all steels are brittle. The question is, at what temperature. Pure iron is ductile at room temperature, but cool it to cryogenic temperatures and it becomes as brittle as glass. Likewise, brittle tool steel can be made ductile by heating. For most metals, there is a temperature that marks a fairly sharp region of transition from ductile to brittle behavior. This is called the ductile-to-brittle (DTB) temperature and is found through what is known as the Charpy impact test. In this test, the energy necessary to fracture a test specimen with a given geometry is measured at a number of different temperatures. The fracture energy decreases with temperature, but at the DTB transition there will be a precipitous drop.

By some estimates the DTB temperature of the steel in the *Titanic* was near 20° C. Not so low that the shipbuilders noticed, but in the –2° C. seawater the day of the collision, the steel at and below the waterline was well below the DTB temperature. On collision with the iceberg, instead of consuming the energy of the impact by rupturing ductile metal, thereby reducing the damage to one or two watertight compartments, only small amounts of energy were used to fracture the brittle steel. The energy of the impact was not consumed until the fractures had propagated along one hundred meters of hull. A little less than three hours later, the ship would sink with the loss of more than 1,500 lives.

While in human terms the sinking of the *Titanic* was a monumental disaster, more critical to the fight to control the problem of fracture was the failure of World War II Liberty Ships. At this time, welding was replacing riveting as the preferred assembly method. Arc welding allowed ships to be constructed at an unprecedented pace, enabling the building of 2,710 Liberty Ships over the course of the war. However, the heat of welding also generated hydrogen, which was rapidly incorporated into the joints. There, hydrogen atoms worked in concert with atoms of sulfur and phosphorus to embrittle the ships. In high seas and low air temperatures, they would literally break in half. The solution was to wrap the Liberty Ships with a steel band just above the waterline, preventing cracks from propagating. A crack stopped growing upon encountering these steel girdles, hence their name, "crack arrestors." Before the crack arrestors were installed, 140 ships a month were lost from all causes; afterwards, the losses were reduced to twenty ships a month.

After the war, George Irwin would use the data collected from the failures of the Liberty Ships to fully describe *when* things break. Working at the Naval Research Laboratories in Washington, D.C., he introduced the concept of a critical stress intensity factor. With this, he was able to clarify crack growth and provide designers with the information they needed to understand when cracks would become unstable and run. The discipline Irwin developed is called fracture mechanics, and is now the foundation for the design of all structures. On that January day of 1979, I knew nothing about Liberty Ships, Irwin, or fracture mechanics. However, I was beginning to realize that embrittlement was the problem to which the answer of *why* things break would provide the solution.

As Greg transitioned from the world of sulfur and phosphorus back to the here and now, he inspected my appointment schedule, and noticed that the next meeting was with Professor Keith Johnson. This seemed to please Greg. He knew that any successful investigation of embrittlement and bonding would require Professor Johnson's participation. Then Mike Wargo arrived to escort me to the next meeting.

I had no idea what to expect. Both Mike and Greg had

spoken of Keith in almost reverent tones. Not in the manner a devout Catholic would assume in talking about the Pope, but rather as a primitive tribe might talk about its shaman, an air of mystery mixed with respect. The mystery surrounding Keith came from the same place as that of a shaman's, in that very few, if any, of the materials science faculty knew precisely what it was that Keith actually did. Oh, they could put words to what he did, but they had no visceral understanding.

Here was a faculty all engaged in essentially the same activity. They sought to design materials with novel properties by manipulating things that could be seen and measured. All of them, that is, except Keith, who saw the properties of materials as arising from the charge density, which cannot be inspected even with the most powerful microscope. Because the charge density is experimentally inaccessible, it must be investigated *theoretically*, solving the equations of quantum mechanics, which govern the motions of things too small to see. If Keith had been a faculty member of a chemistry department, and had restricted his research to small molecules, he would not have been unusual. In many chemistry departments, faculty were engaged in just this type of quantum chemical research. If Keith had been a faculty member of a physics department, and had restricted his research to properties arising from crystalline structure, such as phase stability, again, he would have been in a more typical setting. However, Keith was a senior member of the Materials Science Department at MIT, where the objective was to make new and improved materials. He used tools unfamiliar to materials science, and applied the results differently from the ways chemists and physicists would have. He was therefore in a unique position, intellectually isolated from most of his colleagues, an environment that he actually seemed to enjoy.

Upon entering Keith's office, one could not help noticing the sound system. There were at least four massive speakers, shelves of electronics, and records, shelves of records, extending from one end of the room to the other. Obviously, Keith's passion was music—classical music. I had expected that we would talk about quantum mechanics, but instead we talked about music. Keith told me about his equipment and then said that he was experimenting with "acoustic holography." He sat me in the middle of the room with my eyes closed while he played a couple of selections from the Brandenburg Concertos. I was to point to where the sounds of the instruments appeared to originate. It was amazing; it sounded as if I were sitting right in front of an orchestra. Just as one of my favorite movements was to begin, there was a knock at Keith's door. It was Professor Morris Cohen's assistant, Marge. Apparently Professor Cohen wanted to talk to me. Keith wrapped the meeting up quickly. In five minutes I had filled him in on my background and my interests. Keith's comment was something along the lines that the fracture problem was interesting and he would be delighted to work on it. That was it. He sent me out the door and in the direction of Professor Cohen's office.

Marge introduced me to Professor Cohen. Even before he spoke, there was something in his eyes that communicated intellect. In years to come, I would have occasion to observe Professor Cohen as a scientist and an educator. He is an amazing person. Interestingly, his most impressive virtue was that he simply did not dwell on ideas that were wrong. When most scientists spot a flaw in logic or experiment, their first response is to humiliate the poor sap who made the mistake. Not Professor Cohen. Regardless of the speaker or the subject matter, he always found the part of the presentation that was right or deserving of more investigation. On this occasion, though, he talked about the beauty of science and the importance of viewing the same phenomena from different perspectives. Finally he spoke of Keith as a real asset to the department, bringing a unique point of view to the materials sciences. That was it, a brief but memorable meeting with one of the true icons of the discipline.

Professor Cohen was right; Keith provided important tools to look at materials in new ways. How Keith came to the Materials Science Department of MIT is an important part of the story of fracture. It begins in 1923 as John Slater received his Ph.D. in physics from Harvard, where he worked under Percy Bridgman. Bridgman was a pioneer in the experimental study of substances at high pressure, and later would receive a Nobel Prize for this work. Slater's research concerned the compressibility of simple crystals such as common table salt. His research was experimental. Simply stated, he studied how much force it took to push atoms together and pull them apart. During these investigations, he was troubled by the fact that there were no theories that could predict what he was likely to observe. He resolved to do his best to help in the development of such theories.

Though scientists of Slater's day might disagree with me, the forces that held atoms together and kept them apart were as little understood in 1923, when they were called "bonds," as they had been 2,500 years earlier, when Empedocles had portrayed them as "love" and "strife." While in 1903 atoms were discovered to consist of a massive positively charged nucleus surrounded by negatively charged electrons, no one could explain why this structure was stable, let alone how these atoms could come together and form molecules and solids. The known laws of physics predicted that electrons should spiral toward the nucleus, all the while emitting radiation. Evidently there were unknown laws of nature that permitted the existence of atoms and molecules, and at the outset of the twentieth century, the search for these laws was in full swing.

By 1923 a great deal of suggestive experimental evidence concerning these laws had been amassed. Still, there was no overarching theory to explain these observations, but only bits and pieces, which together were being called the "quantum theory." The most influential and senior scientists of the quantum theory were in Europe. Among these was Niels Bohr, and so it was to Europe with the express intent of spending time working with Bohr that Slater went in 1923.

Laws of nature cannot be derived through clever applications of mathematics; they simply exist, and must be intuited. Scientists of Bohr and Slater's time suspected that these hidden laws would be different from the classical laws in several respects. First, classical theory assumes that a particle changes its energy continuously. For example, a car accelerating from zero to sixty passes through every speed between zero and sixty. Under the new laws, this would not always be the case. Some things would change energy in steps, as if an accelerating car started from zero and jumped to sixty without passing through an intermediate speed. Energy appeared to come in packages called quanta, from which the name "quantum theory" derives. There was a second, more mysterious aspect to the laws. Classically, things come in two different forms: particles and waves. A baseball is a particle and sound is a wave. We never observe a baseball wave or a sound particle. Under the new laws, it appeared that sometimes a thing was a wave and sometimes a particle. This odd characteristic of matter and energy was named wave-particle duality, and no one knew how to incorporate it into a new theory.

While working at the Cavendish Laboratory at Cambridge University, Slater had an amazing insight. He was attempting to explain why excited atoms, as they decayed, did not always emit light with the same amount of energy. He realized that this could be readily explained by assuming the emitted light was a particle. The energies of the emitted particles, called photons, were determined by a probability distribution. Following this line of reasoning, Slater recognized that the wavelike properties of matter and energy were a consequence of the wavelike properties of the laws of probability.

In December 1923, while in Copenhagen, Slater shared

his insights with Niels Bohr and Hans Kramers. Both were enthusiastic, but preferred to think of the probabilistic nature of matter and energy as a kind of "mathematical trick" that was useful in extracting the right results from the existing bits and pieces of quantum theory. One of these bits was the Bohr model describing the excitations of electrons in the hydrogen atom. Bohr, Kramers, and Slater collaborated on a paper explaining the energy distribution of emitted light from excited hydrogen atoms. Much to Slater's consternation, the paper adopted Bohr's interpretation for the wave-particle duality of light. As Slater recounts the episode in his autobiography, the conflict led to a great "coolness" between himself and Bohr, which was never completely removed. Slater returned to a faculty position at Harvard late in 1924.

Why an apparently legitimate scientific difference of opinion should lead to such "coolness" on Slater's part can be inferred from some of his writings. It is possible that Slater felt he would have been able to extend his insights and develop the full theoretical basis of quantum mechanics if he had not had the "antagonism of Bohr" to contend with. The use of the words "coolness" and "antagonism" suggests that the Slater-Bohr dispute was not a polite disagreement but rather a full-blown scientific struggle. Nevertheless, two years later, in 1926, the young physicist Erwin Schrödinger published the first papers to elucidate the physical laws governing electrons, atoms, and molecules. These laws constituted the foundation for a new science of quantum mechanics. At its heart is the Schrödinger wave equation, which explains wave-particle duality in terms of the probability of making measurements. For his part in the development of quantum mechanics, Schrödinger would be awarded the Nobel Prize in physics.

After the discovery of the laws of quantum mechanics embodied in the Schrödinger wave equation came the real work—using these laws and equations to uncover the hidden secrets of the molecular world. It was to this task that Slater would devote the rest of his life. He appreciated that the great technological advantages that could flow from quantum theory required that it be accessible to chemists and metallurgists. To this end Slater sought to construct useful yet quantum-mechanically correct representations of the chemical bond. During the years 1927 to 1931 he and his close friend, the chemist Linus Pauling, independently developed the "directed valence" representation of chemical bonding. This became the standard approach for rationalizing chemical reactivity and molecular structure. It is taught today in all general and organic chemistry classes. This theory reportedly convinced the great chemist Gilbert Lewis of the correctness and usefulness of quantum theory. Pauling would be awarded the Nobel Prize in chemistry for his contributions to the theory of directed valence.

In 1930, Slater would move down the road from Harvard to become the head of the Physics Department at MIT, where he would remain until 1964. During that time he brought together a truly exceptional group of scientists who would make major contributions in diverse areas of physics. However, Slater's passion was the advancement of quantum theory toward an understanding of molecules and solids, and in this pursuit he worked tirelessly. Although a complete enumeration of his accomplishments is beyond the scope of this book, a few highlights are worth recounting. Some are essential to understanding the unique environment in the Materials Science Department of MIT in 1979.

In 1938, a very bright undergraduate by the name of Richard Feynman turned up in one of Slater's classes, looking for a project. Slater challenged him to prove what, as far as Slater knew, had been only surmised, that the forces on atoms could be calculated from a knowledge of the electron density of a given volume—that is, the number of electrons in it. The young Feynman did prove the surmise correct. Only later was it discovered that Hans Hellmann, in Germany, had proved the same theory three years earlier. Consequently, their effort would be known as the Hellmann-Feynman theory. It provided for a rigorous understanding of bond formation. Feynman would become a great physicist and educator and would be awarded the Nobel Prize in Physics for later work unrelated to the Hellmann-Feynman theory.

Slater seemed to develop close relations with people who applied quantum theory to make and understand things. Among these was another of Slater's students, William Shockley. Shockley was interested in what is called band theory. This is simply a theory that describes the bonding in crystals. It is a particularly useful place to start when trying to account for the movement of electrons through crystals, what we call electrical conduction. Shockley would later use band theory in the invention of the transistor while working at Bell Laboratories. For this, he and his co-inventors would win the Nobel Prize.

With the development of digital computers in the 1950s and 1960s, Slater saw another opportunity to make quantum mechanics accessible and useful. He was, in fact, the founder of the disciplines that use computers to solve quantum mechanical equations governing the properties of crystals. This represents a significant portion of what is now called computational solid-state physics. It is in this area that we see Slater's true legacy. Many, perhaps a majority, of those working in the field of computational solid-state physics today are the intellectual grandchildren of Slater. His research group at MIT dominated the field. For those wishing to work in this area, a pilgrimage to MIT was one way of establishing legitimacy. And so it was that many of those who became the leaders of computational physics in the 1970s did postdoctoral research under Slater.

Getting along with Slater was not an easy matter, and many of his graduate students, postdoctoral associates, and colleagues did not. My suspicion is that he was dogmatic in his desire to make quantum mechanics useful, and in this regard he relied heavily on intuition. I believe he had little tolerance for those who would plod through a calculation and then garner little useful information, other than that this or that *had* to be done to make the calculation more accurate. Making calculations accurate has become a focus of computational physics. Since Slater didn't share this particular value, he was isolated from the practitioners of the very discipline he founded. The level of resentment he generated in the Physics Department at MIT is evident. Slater retired from MIT in 1964. He was head of the Physics Department for nearly thirty years, yet unlike all other department heads, there is not a room, an endowed chair, or a memorial lecture named for him. In fact, I was unable to find even a picture of Slater in the MIT Museum. It was as if he did not exist.

In the years preceding his retirement from MIT, Slater was asked to sit on a panel to investigate whether the Atomic Energy Commission should support university laboratories doing research in the field of properties of materials. Also on this panel was Morris Cohen, the same Morris Cohen who would later speak to me about the beauty and connectedness of science. Slater and Cohen discovered that they had similar views about the nature of interdisciplinary research, and they took it upon themselves to organize a materials program at MIT. The result was the Vannevar Bush building housing the Materials Program at MIT, dedicated in October 1965.

In September 1964, Slater retired from MIT and moved to Gainesville, Florida, where he established the quantum theory project at the University of Florida. In 1965, Keith Johnson began his postdoctoral year in Florida, working in Slater's group. Slater later described Keith as "an able young worker, equally interested in physics and chemistry." Slater was obviously impressed with him, and must have played a significant part in helping him to secure a position at MIT as an associate professor in 1966, not in the Physics Department, where Slater no longer held sway, but in the Department of Metallurgy and Materials Science.

Slater died in 1976. In that year he had been selected to win the Nobel Prize for his quantum mechanical investigations of magnetism, but he died before the announcement was made. Because the Nobel Prize is never presented posthumously, he did not receive the award. In 1999, the Nobel Prize for Chemistry was awarded to the physicist Walter Kohn for work in density functional theory, another area in which Slater was a major player.

So it was into that environment that I drifted because of accident and circumstance in 1979. There I was about to become a graduate student at the only university in the world where the expertise and interest resides to study why things break. And even though the field was my passion, I really only stumbled upon MIT because my girlfriend liked Boston and my kayak had broken a year earlier. Cheryl actually did agree to move to Boston only because she had such a good time on our interview trip. Our most delightful memory of this trip revolves around two policemen who gave us a tour because I had forgotten my coat. The world is strange indeed.

SHOCKING, SIMPLY SHOCKING

CHAPTER 5

In my opinion, fracture mechanics is one of the great triumphs of science and engineering. At the turn of the millennium, numerous TV shows and magazine articles detailed the 100 greatest accomplishments of the last thousand years. It was hard to dispute the majority of achievements that made the list—things like Gutenberg's invention of movable type. I did, however, find myself wondering about some significant milestones that were left off. I do not know of a single list that included the discovery of fracture mechanics as a great accomplishment. Sure, you could argue that my area of specialty made me notice what I consider to be a glaring absence, but bear with me here, because this discovery pushed our society to make a quantum leap, wherein the expectations of the things we build and the engineers who build them changed forever.

Now, "quantum leap" is a perfectly good phrase, which I enjoy using because my field is quantum mechanics. It distresses me, however, that few people really understand what the phrase means. Several years ago, a radio talk-show host spent half an hour explaining to all in the range of his voice that the TV series *Quantum Leap* was misnamed and the phrase in general was misused. After all, he argued, a quantum of energy is a tiny amount—so a quantum leap would refer to a tiny leap. Clearly, he went on to add, the term is used in reference to large changes and is therefore misused. Though a talk-show host can be forgiven his ignorance, some of my esteemed colleagues have also espoused the very same argument. It is not quite as forgivable in their case, because, simply put, it is wrong.

First, a quantum of energy *need not* be tiny. Though the quantum phenomena we typically study involve tiny amounts of energy, there are also very large quanta. I think the misunderstanding has its roots in the dictionary definition of quanta. A dictionary definition of *quantum* is "the smallest amount of a physical quantity that can exist independently, especially a discrete quantity of electromagnetic radiation." This is a good definition. However, the smallest amount of a thing need not necessarily be small. In some theories, the Big Bang, which gave birth to the universe, was a quantum transition. In these theories the smallest amount of the universe that can exist independently is the whole universe. Not something small at all.

The term *quantum* makes no reference to the size of the energy involved—it can be very tiny or unimaginably large. What the "quantum" of "quantum leap" implies is that something went from one state to another without passing through intermediate states. First there was no universe and then, all of a sudden, there it was. There were no intermediate states where there were fractions of a universe.

Returning to the effects of fracture mechanics, on Friday, April 19, 1912, the Waldorf-Astoria Hotel in New York City was the location of hearings into the circumstances surrounding the sinking of the *Titanic,* just four days earlier. Senator William Smith of Michigan chaired the investigation of the Committee on Commerce, which was officially involved in waterway navigation and safety. As were many, Senator Smith was aghast at the senseless loss of life and wished to use the findings of the committee's investigation to draft "an international agreement to secure the protection of sea traffic."

For slightly more than two weeks, first in New York and then in Washington, D.C., the committee heard testimony from the survivors of the tragedy, uncovering numerous examples of gross negligence on the part of the *Titanic*'s captain and crew. Most egregious of these was the captain's failure to act in accordance with the warnings of the crew of the ship *California,* who had advised the *Titanic* of iceberg sightings. The *California,* in fact, passed the night at rest on the calm seas of the North Atlantic while the *Titanic* actually increased speed. The White Star Line, which owned the *Titanic,* was also negligent, having outfitted the ship with too few lifeboats to accommodate all the passengers and no binoculars for the lookouts—though these were not requirements for licensure by the British Board of Trade. Under the laws of 1912, none of these rose to the level of legal negligence, and all that the committee could do was to write new law to prevent such a tragedy in the future.

Significantly, little or no attention was given to the materials and construction of the *Titanic.* It was believed that even though the human toll could have been reduced through proper action, the sinking itself was seen as an act of God, no more avoidable than the toppling of the Tower of Babel—a consequence of man's pride in materialism and technology. That a ship could be designed to withstand an impact with an iceberg seemed amazing hubris in light of recent events. This reaction stands in marked contrast to an event of my lifetime in which pride and the improper reliance on technology overshadowed common sense.

For me that day unfolded as I drove to work at Los Alamos National Laboratory in New Mexico, my first job after leaving MIT. It was a spectacular January day in 1986. Though the lab is located on a mesa nearly 7,500 feet above sea level, the winter of 1985–86 was unusually mild. I found the almost tropical weather disappointing. Since moving to Boston in 1979, I had all but forsaken skiing. The icy slopes of New England provided an unsatisfactory substitute for the champagne powder of the Rocky Mountains. And though the ski season in New Mexico was shorter than that of Colorado, I was eager to ski once again on something other than an inclined ice rink. Still, despite the lack of snow, the spectacular cloudless blue sky of New Mexico was a huge improvement over Boston's dreary gray winters.

While New Mexico enjoyed extraordinarily warm winter temperatures, normally moderate Florida was experiencing an unusual cold spell. On the afternoon of January 27, the temperature began to plummet at Cape Canaveral, where the space shuttle *Challenger* was awaiting launch. Three times the launch of the *Challenger* had been scrubbed, twice because of adverse weather conditions and then again on the morning of the twenty-seventh, when the close-out crew found a stripped screw preventing them from removing the handle from the space shuttle's door. After five farcical hours, during which a succession of attempts to use a portable electric drill failed because charged batteries could not be found, a hacksaw was employed and the handle was finally removed. Then, after all that, developing crosswinds forced the third cancellation of the launch.

This was no ordinary launch. On board was Christa McAuliffe, a schoolteacher from New England who had beaten out 11,000 other educators to win a place on this shuttle mission. Her presence was part of a plan by the Reagan administration to boost flagging interest in the space program. I am a cynic and have little regard for the lip service given to fundamental research by politicians, from either party. For this administration the real interest in the manned space program derived from the president's commitment to the space-based defense system colloquially referred to as "Star Wars." One element of this program required an orbiting command-and-control facility. The space shuttle was, in the view of many in the administration, a key element to building that part of the system. Still, setting my usual cynicism aside for the moment, whatever the motivation, Christa, as the "ordinary person" venturing into space, had indeed generated interest. NASA was at center stage, and with the interest came impatience as the world waited for the launch of the schoolteacher.

The debacle of January 27 did not engender the kind of publicity NASA had hoped for. The image of "rocket scientists" taking five hours to remove a stripped screw was served up for the amusement of the viewing audience by the late-night talk-show hosts. Undoubtedly the managers at NASA wished to avoid providing fodder for a repeat performance. So the dropping temperature was probably regarded as just another opportunity for embarrassment.

As ice began to form on the *Challenger*, there was an expectation that the launch would again be canceled. Ice falling from the shuttle and damaging the external fuel tanks was put forward as a possibility. Hence I was amazed as I began my commute to Los Alamos on the morning of the twenty-eighth and flipped on the car radio to discover that the countdown was well under way—T minus twenty-eight minutes.

I lived in the valley 1,500 feet below and about twenty miles from the lab. Though the commute took about half an hour, the scenery was astonishing and every day I looked forward to the drive. After crossing the Rio Grande—T minus eighteen minutes—I began the climb to the top of the mesa where the lab sits. The mesa is the ashy remnant of a tremendous volcanic explosion millions of years ago.

Over time the ash compressed into a dusky pink rock called volcanic tuff. Through the action of wind and water, the mesa was fashioned into a series of gnarled fingers that reach for the waters of the Rio Grande. On this morning, the long pink fingers of the mesa stood in stark relief against the cloudless deep blue New Mexican sky—T minus thirteen minutes.

After crossing the Rio Grande, one has a choice, to follow the mesa's index finger to the top and then turn south and take a bridge connecting the index and middle fingers, or to turn at the bottom of the cliffs and climb the mesa via the middle finger. As usual, I turned south at the bottom of the mesa—T minus ten minutes. At the tip of the middle finger and just out of sight over a small rise is an archaeological site. The soft tuff is easily carved, and 800 years ago Native Americans turned this to advantage, fashioning dwellings in the ancient volcanic stone. The only indication of this site is a nondescript sign, easily overlooked by the passing drivers on their way to the main entrance and the more extensive cliff dwellings and ruins of Bandelier National Monument. Here I turned away from the ruins, taking the six-mile road up the mesa's middle finger—T minus eight minutes. Climbing through ponderosa pines, past the practice range where the security forces defending the lab sharpen their marksmanship skills, one arrives at the top of the mesa just a few hundred feet from the largest of the lab's many technical areas, or TAs. It was here that I worked—T minus one minute. Pulling into the parking lot and looking east, the view is unobstructed. Sandia Peak, the southernmost of the Rocky Mountains, is almost always visible, as is Mount Blanche, far to the north in Colorado—liftoff.

I sat in the car listening to the communications between flight control and the shuttle. Looking east, I half expected to see it emerge from behind Sandia Peak. Over the radio, one of the astronauts said, "Thirty-five thousand going through one point five" indicating that the shuttle had just passed through 35,000 feet and a velocity one and a half times the speed of sound, about 1,500 feet per second. Then, from another astronaut, "Roger go at throttle up."

Then a long silence, followed by, "Can it . . . Can it? Oh my God, can *Challenger* have exploded? Oh no! What on Earth has happened?"

For four hours I sat in my car listening to the reports, all the time waiting to be told what had happened, what had broken and caused this accident. Had falling ice punctured the exterior fuel tank? By afternoon it was clear that answers were not going to be forthcoming, at least in the near term. An investigation would follow, and we would have to wait for answers. While the rest of the world waited, there were several people working for one of NASA's contractors who were already well aware of what had happened; they had in fact *predicted* it.

One of them was Roger Boisjoly, a senior engineer at Morton-Thiokol. Another was Bob Ebeling, his manager. Morton-Thiokol manufactured the 126-foot solid-fuel rocket boosters generating up to 75 percent of the thrust needed to catapult the shuttle into orbit. Each booster is filled with about a million pounds of solid propellant—much like the propellant used in those Estes rockets many of us played with as kids—before being moved to Cape Canaveral by rail. Because no railway can transport an object the size of a ten-story office building, Morton-Thiokol built the boosters in sections and shipped them to the Cape, where they were assembled and strapped to the external liquid fuel tank of the shuttle. When ignited, each section of the booster expands in response to the tremendous heat and pressure of the exhaust gases produced through the burning of the solid rocket fuel. During this expansion, something has to seal the joints between the sections of the booster, if hot gases are to be prevented from pouring through the joints. Two quarter-inch rubber O-rings were used for this purpose. They were to expand rapidly, sealing any gaps between the sections.

For about six months before the *Challenger* explosion, Roger Boisjoly had been concerned that as the O-rings became cold they would expand more slowly than required to seal the booster sections. If this were to occur, the escaping gases could burn into the external fuel tank containing the liquid hydrogen and oxygen, causing it to explode. Boisjoly's fears had been growing for about a year, after he had inspected the O-rings from the spent boosters of a previous mission. On this particular occasion the shuttle had been launched in quite moderate temperatures—a usually comfortable 53 degrees. To Boisjoly's astonishment, surging hot gases had scorched the primary O-ring. Fortunately the secondary O-ring had held, possibly preventing an earlier tragedy.

As expected, NASA considered the performance of the O-rings to be a critical issue. A task force was created at

Morton-Thiokol to look into the problem. Unsatisfied with the pace of the investigation, however, Boisjoly sent a memo to Thiokol's vice president, articulating the serious nature of the problem. He wrote, "It is my honest and very real fear that if we do not take immediate action . . . we stand in jeopardy of losing a flight, along with all the launch pad facilities. The result would be a catastrophe of the highest order—loss of human life."

On the afternoon of the twenty-seventh, the temperature began to drop at the Cape. The prediction for the next day's launch was an unprecedented 23 degrees—a full 30 degrees colder than the flight that had triggered Boisjoly's initial fears. Having no experience with such matters, the launch team called its contractors to ask if they had any concerns. The engineers at Morton-Thiokol gave an unqualified and emphatic "yes" in response.

A nearly six-hour conference call between NASA rocket specialists at Marshall space flight center and the engineers and management at Morton-Thiokol ensued.

For five hours Boisjoly, Ebeling, and their colleagues argued that the launch should be scrubbed. The men at Marshall were unconvinced, believing that the Thiokol engineers were giving voice to intuition rather than sound science. After five hours the NASA representatives indicated that the launch would not proceed without contractor approval, but they believed Thiokol's recommendation was without foundation.

In the world of science and engineering, having a reputation for letting your gut drive your decisions, instead of your head, is about as bad as it gets. Under the pressure of this accusation, Thiokol management buckled. The four senior managers ignored the best advice of their engineers and made a "management decision" changing Thiokol's recommendation to "go for launch."

Letting my cynicism return for the moment, I have to believe that this is *exactly* what NASA wanted. NASA needed to show the president that neither rain nor ice nor gloom of night could stay them from launch. After all, the command and control for Star Wars would have to fly in all conditions, as the Soviets might not be kind enough to wait for clear weather to launch an attack. Besides, the State of the Union Address was slated for the evening of the twenty-eighth, and having Christa orbiting the Earth, some at NASA must have reasoned, might capture some recognition in the president's speech. And so, the stage was set.

As I was driving to Los Alamos, and Christa McAuliffe was at the Cape eagerly anticipating the next few days, Boisjoly was walking past the conference room at Thiokol's Utah facility, where Ebeling was watching the launch on TV. After some persuasion the two engineers watched the last few minutes of the countdown together. At T minus five seconds, when the solid rocket boosters fired, I was looking across the high desert of New Mexico. Christa McAuliffe's heart must have been pounding with excitement, while she was unaware that puffs of black smoke were emerging from the joints of the starboard solid rocket booster. In Utah, Boisjoly and Ebeling joined hands and braced for an explosion. Yet there was none, and both men breathed a sigh of relief. Then, seventy-three seconds into the flight, it happened, a new cloud burst into being in the sky above Cape

Canaveral—remnants of *Challenger's* liquid fuel. In New Mexico, I sat in my car waiting to learn what had caused the explosion; and in Utah, Boisjoly, who already knew, sat in his office facing a wall while trying to hold back his emotions.

As with the *Titanic,* a governmental commission would be appointed to investigate the cause of the *Challenger* explosion. Unlike the *Titanic,* here the government had a major stake in the conclusions drawn by the investigators. *Challenger* was not the property of a private enterprise and licensed by another government; it flew and exploded as part of a governmental program. So who, from the government, could be expected to investigate the tragedy impartially?

Only a few days after *Challenger* exploded, preparations were under way to form a presidential commission to investigate and collect information. William Rogers, secretary of state in the Nixon administration, chaired the commission. Neil Armstrong, the first man on the moon, was the vice-chairman. Sally Ride, the first American woman in space, was also on the commission along with others, including an air force general, Donald Kutyna, and a physicist and Nobel laureate from Cal Tech, Richard Feynman—the same Richard Feynman who had worked with John Slater at MIT.

In all the commission, Feynman was probably the member with the least at risk if the final report was critical of NASA. He was also about as close as a scientist can get to being a pop icon. His thoroughly entertaining autobiography, *Surely You're Joking, Mr. Feynman,* was widely read, even by people with little or no interest in science. As a result, other members of the commission, particularly General Kutyna, funneled information to the public through Feynman. Notably, Kutyna, who was probably one of those who suspected the failure of the O-rings had caused the explosion, started Feynman thinking along these same lines. In a rather dramatic piece of theater, during open hearings, Feynman plunged samples of O-ring material into a glass of ice water to show that they would lose pliability. Though Rogers quickly refocused the hearings, the information was out. That night the "Feynman demonstration" was replayed on the evening news, and by the next day the newspapers seemed to have made the connection between the cold weather, O-rings, and the *Challenger* explosion.

The final report of the commission included what was apparently an unwelcome supplement by Feynman. In it he gave a scathing analysis of NASA's risk assessment. Before the *Challenger* tragedy, NASA estimated the probability of catastrophic mission failure at upwards of 100,000 to one. Such estimates enraged Feynman. Because there are thousands of critical components in a shuttle, many of which have no redundant backup, this estimate required failure probabilities of individual components to be on the order of billions to one, an absolutely unattainable goal. Using more reasonable probabilities, Feynman estimated catastrophic mission failure to be less than 100 to one. On February 1, 2003, during the 113th shuttle mission, the *Columbia* broke up on reentry, killing all seven astronauts aboard.

To me the similarities between the *Titanic* and *Challenger* tragedies are uncanny. Both disasters could have been prevented if those in charge had heeded the warnings of those who knew. In both cases, materials failed due to thermal effects. For the *Titanic*, the steel of her hull was below its ductile-to-brittle transition temperature; and for the *Challenger*, the rubber of the O-rings lost pliability in subfreezing temperatures. And both tragedies provoked a worldwide discussion about the appropriate role for technology. However, there were also differences. The most significant of these dealt with remedies. In the case of *Challenger*, the attitude from the outset was "Identify what failed and fix it." For the *Titanic*, the attitude was "Avoid similar circumstances in the future." The more modern attitude reflects the growing belief that the world and the things we build can be made safe, whereas the earlier attitude reflects the belief that the world is inherently hazardous and risks are better avoided than addressed. I consider this change in attitude to be a quantum change, driven by technology in general and fracture mechanics in particular.

Jump forward seven years. On Christmas Eve, 1993, Patricia Anderson, her four children, and a family friend were on their way home from church. As they waited for a traffic light to change, a drunk driver traveling between 49 and 70 mph plowed into the rear of their 1979 Chevy Malibu. The Chevy's rear-mounted gas tank ruptured and exploded. Flames engulfed the passenger compartment, seriously injuring and permanently disfiguring all six occupants. The only non-fire-related injury was a broken leg suffered by one of the children. The drunk driver—whose blood alcohol level was 0.2 percent—sustained only minor injuries.

What an incredibly tragic accident. Over the next six

years Patricia Anderson sought to attribute liability not to the drunk driver, but to General Motors for manufacturing her "unsafe" fourteen-year-old car. On July 9, 1999, a superior court jury in Los Angeles awarded Anderson $4.9 billion. That's not a misprint—$4.9 *billion*. The award included $107 million in compensatory damages and $4.8 billion in punitive damages. What on Earth had GM done to merit this judgment?

Anderson's legal case was built on the theory that the Malibu's fuel system was inherently unsafe, and that GM knew this and could have designed a safer system but had decided not to in order to boost profits. What a change. When the *Titanic* sank, no one suggested that the engineers of the day should have built a ship more resistant to icebergs. Of course, at the time, without the discipline of fracture mechanics, the engineers could not do this—unless they had built many *Titanics* and rammed them intentionally into icebergs until they found a safer design. Now the attitude appears to be "Because it is possible to predict when something will break, it must be designed so that it will not, regardless of the cost." Actually, this view is not quite complete. The plaintiff argued that it would have cost GM only a few dollars per car to build a safer fuel system. In support of this position, the plaintiff's lawyers produced a memo written by GM engineer Ed Ivey. In this 1973 memo, Ivey reported on a cost-benefit analysis which concluded that the costs to GM from fire-related injuries amounted to $2.40 per car. However, the cost of upgrading to more impact-resistant fuel systems would cost between four dollars and twelve dollars per car.

On the other side of the coin, GM's defense rested in convincing the jury that no fuel tank could withstand the impact of the collision. The defense argued that the drunk driver was traveling over 65 mph, and the Malibu's fuel tank was already designed to withstand a collision of 50 mph.

Dueling experts argued for days. Was the drunk driver going 49 mph, as plaintiffs claimed, or 65 as the defense contended. I, for one, believe 49 mph is a bit too convenient. No matter what, experts tend to "see" the facts from the perspective of those who are paying them. It is not as if they purposely distort their opinions, they just weight things in favor of their clients. This is why, though I don't make them often, nine out of ten arithmetic errors in my checkbook are in my favor. I am inclined to split the difference—the drunk driver was probably doing between 55 and 60 mph. The jury didn't see it this way and found for the plaintiff.

As a scientist and an engineer I find both the arguments of the plaintiff and the defense a little weak, because both ignore the design process. Most disturbing, however, was GM's apparent belief that it could not rely on a jury to come to an equitable decision by just laying it all out on the table. Of course GM could design a fuel tank that would survive a 50-, 60-, or even 70-mph collision. Anyone who has seen a collision at a NASCAR race knows this. These cars collide at speeds over 200 mph and, more often than not, the drivers walk away. However, a racing stock-car costs many times more than most people can afford to pay for their family transportation. Consequently GM and every other automotive manufacturer must strike a compromise between a host of design attributes including affordability, profitability, style, performance, reliability, and *safety*.

What the jury interpreted as GM placing profit above human life could be seen as an attempt to make the safest car possible for a given cost. Sure, it would cost two dollars to ten dollars per car to upgrade the fuel system, but is this the best use of that money? Maybe those same funds should be used to redesign head restraints, or improve the energy dissipation of the overall structure. Attention paid to these safety features would increase the likelihood of survival in a high-speed rear impact. Curiously, the fact that, other than the terrible burns, the most serious injury was a broken leg was not seen as a triumph of engineering but as justification of an expectation that the things we build should never break. How is the designer to make the decision whether to use ten dollars to upgrade head restraints or to put it into fuel systems? I know of no way other than to compare relative costs. If the cost to GM from inadequate head restraints is greater than the cost of fuel systems, the engineer is justified in making the first priority head restraints.

We have two choices. The first is to make safety our primary concern, as should have been done for the *Challenger* astronauts. With this priority, we cannot question cost or be concerned with delays. The second choice is to acknowledge that the world is a risky place and expecting technology to eliminate that risk is unreasonable. It appears, however, that we are becoming increasingly impatient with our technology and less willing to accept the risks associated with this impatience. We are making designers and engineers the scapegoats for our inability to both have technology and use it without risk, too.

It was with a call to my university office in 1995 that I first began to recognize how the companies designing the products of our world are increasingly being placed in an untenable position. As happens every so often, this was an inquiry concerning my availability to act as an expert witness in a lawsuit. Often, whether or not something *should* have broken, or *how* it broke, is of central importance in civil lawsuits. In this instance, how and when a cast-iron pump broke was important in fixing liability for injuries suffered during a refinery fire.

To the best of my recollection, these are the facts surrounding this case:

On a cool evening in the early part of June 1992, workers at the Frontier refinery in Cheyenne, Wyoming, were attempting to pump waste product into a "slop system." Here it was to be stored for recycling and recovery of crude oil in the future. However, things were not going well because the waste product was too viscous to pump, apparently owing to the nearly freezing temperature. At this point someone came up with a bright idea: heat everything up by pumping warm hydrocarbons into the slop system. Accordingly, roughly room-temperature hydrocarbons were diverted by means of a centrifugal pump manufactured by Gorman-Rupp into one of the very large storage tanks of the slop system. Shortly thereafter, there was an explosion and fire that killed one worker, Timothy Price, and severely burned four contractors who were working in the area of the slop system. Suits were filed, which Frontier and its liability insurers ultimately settled for approximately $19 million. This all occurred well before I was contacted.

My involvement centered on determining what had caused the fire. And as with all litigation, there were two stories about how the incident had occurred. In the first scenario, the warm hydrocarbons entering the cold tank caused some of its contents to vaporize, increasing the pressure and ultimately rupturing the tank. The fire started at the storage tank. In the second scenario, the cast-iron pump used to divert the warm hydrocarbons into the cold storage tank fractured as a result of thermal shock as its temperature changed suddenly from near freezing to room temperature. The fire started at the broken pump and then spread to the storage tanks, which then exploded. If scenario one was correct, then the refinery and its insurance companies were libel for damages. On the other hand, if scenario two was correct, then, in the opinion of the refinery's lawyers, the pump manufacturer bore some, if not all, of the liability for the accident.

I do not often get involved in litigation, largely because it is often simply impossible to conclude with scientific certainty what caused something to break. The fact that almost every case involves opposing experts just serves to confirm this difficulty. However, as I heard the facts of this case I quickly formed a scientific opinion. Cast iron will not fracture through thermal up-shock—that is, by raising its temperature. Thermal shock causes fracture only when the temperature is suddenly dropped, in a process called down-shock. This I knew from my experience as a child, breaking heated marbles by dropping them into cold water. They never broke while being heated on the cookie sheet. The attorneys who had contacted me represented the pump manufacturer. In my mind there was no question who was on the right side of this issue, and I agreed to serve as an expert for Gorman-Rupp, the pump's manufacturer.

The pump was made from cast iron, remarkably versatile and commonly used because products are so easily fabricated. As the name implies, objects made from cast iron are typically formed by casting. Molten iron, containing approximately two percent carbon, flows well and melts at a lower temperature than pure iron. This makes it especially well suited for casting, an economical method of fabrication. Further, cast iron is corrosion-resistant and rigid, making it an ideal material from which to construct many common products, from engine blocks and frying pans to bathtubs and architectural elements. It is also considered brittle, but this does not imply that it shatters like glass.

When Cheryl and I moved to Boston in 1979, we purchased an old fixer-upper of a home in a Boston suburb. One of our first remodeling tasks was the second-floor bathroom, where we planned to remove an old cast-iron bathtub. This tub must have weighed 300 pounds. Rather than removing it intact, I had the idea of breaking it into several pieces. Cast iron is brittle, so I thought this should not prove a major effort. Using a twenty-pound sledgehammer, I beat on that tub for fifteen minutes without producing so much as a chip. Not to be defeated by a bathtub, I used a hacksaw to cut a one-inch "crack" through the edge of the tub. Now, with each blow of the sledgehammer, the crack extended, eventually separating the tub into two pieces. I repeated the process, reducing each half of the tub to two pieces. The resulting bathtub quarters were removed more easily than the full tub would have been, but in the process, I gained a new appreciation for supposedly *brittle* cast iron.

The refinery, however, was not alleging that the pump had failed from impact but from thermal shock, and up-shock at that. Sudden temperature changes can result in fracture owing to thermal expansion and contraction. Almost everything expands on heating and contracts on cooling. There are a few exceptions. Water, in a very narrow temperature range between 0° and 4° C., contracts on heating, as does one of the stable solid phases of plutonium. But cast iron is typical and expands when heated.

As explained above, in down-shock the object is suddenly cooled, while in up-shock it is suddenly heated. Down-shock caused my heated marbles to break when dropped into cold water. The heated marbles had expanded and when plunged into the cold water began to contract. However, the surfaces of the marbles cooled faster and therefore contracted more rapidly than did their interiors. The surface of the marbles became too small to contain their insides and pulled apart. In the converse, up-shock, a cooled and contracted item would be heated by, say, plunging it into hot water. The surface would expand faster than the interior, producing compressive forces, which cannot cause planes of atoms to come apart. Up-shock cannot cause a normal material to fracture—plutonium, maybe, but cast iron, never.

Think of all of the applications where cast-iron products are subjected to up-shock as part of normal use. A cast-iron engine block does not fracture when, on a cold winter day, it is rapidly heated by several hundred degrees. My grandmother's cast-iron frying pan did not break when placed directly on a gas burner, and a cast-iron bathtub will not fracture when scalding hot water is suddenly sprayed from a shower head onto its surface. However, we are warned to avoid subjecting many of the same items to down-shock. Never is cold water to be run through an overheated engine for fear of "cracking the block." My grandmother reprimanded me for warping her skillet when she caught me running cold water on its heated surface to see the eruption of steam.

Nearly everyone has direct experience with cast iron and its thermal properties. So common are these experiences that I found it hard to understand how the refinery could plausibly argue that the pump had failed from up-shock. They did have a fractured pump, but that was easily explained. No one disputed that the pump had been engulfed in flames. In fact, the zinc bushings around the impeller shaft melted from the intense heat of the fire. Zinc melts at 419° C., indicating that the pump reached at least this temperature during the firefight, where high-pressure water hoses had been used. The cold water from these hoses had cooled the pump rapidly. The resultant down-shock had fractured the pump. In my opinion, the refinery was grossly negligent, and, in an attempt to avoid responsibility, had combed the wreckage of the fire looking for *anything* that could be used to divert attention from its negligence. What the company came up with was a broken pump and a proposition that was scientifically improbable, if not impossible.

Though I was incredulous that the refinery would even

attempt to perpetrate such a hoax, I was even more astonished that they were able to locate an "expert" to support their position. A professor of mechanical engineering at a reputable university had done "calculations" to show that the pump could have broken in the manner asserted by the refinery. The calculations, however, confused thermal shock with thermal stress.

Thermal stress results when two materials that expand at different rates are placed in contact and then heated or cooled. Thermal stress is a common cause of pops and creaks in old houses and in extreme cases has been found to be the true origin of hauntings. During the heat of the day, rafters expand more than foundations. In a well-built house, as the rafters expand they are designed to "roll" smoothly over the bearing walls, but in an old house, friction between the rafters and these walls prevents this smooth motion. Stresses build in the rafters until they overcome the friction, and then with a sudden motion the stresses are relieved. These motions are frequently accompanied by loud pops and creaks.

If the pump had been securely fastened to a base that expanded much less than the cast iron, then heating the pump could cause it to fracture. However, this fracture would originate at the point where the pump had been secured. Both the plaintiff's and defendant's experts pointed to the pump body as the point of fracture initiation. Apparently the defendant's "expert" did not understand the very equations used to make his arguments.

As the case progressed, it just seemed to get more bizarre. Evidently the refinery had not received the pump directly from the manufacturer, but had purchased it used. To the best of my understanding, the plaintiffs were asserting that the pump manufacturer was negligent for making a pump from cast iron, which was likely to fail during up-shock and cause a fire, if used to pump flammable materials. Yet it was the decision of the engineers at the refinery to purchase this particular pump and use it in this particular manner.

From my perspective, which was admittedly myopic, since I only availed myself of information that directly affected my opinions, the whole case hinged on a scientific question: Could a cast-iron pump fracture during up-shock? We said no, they said yes. Such questions of science are best answered by testing them. It would have taken no time at all to cut samples from the fractured pump, cool them in ice water, and subsequently plunge them into room-temperature water. If even one broke, we would be wrong; if not, then the plaintiff's theory would be placed in serious doubt. Why the court should allow months of discovery, hours of depositions, and days of trial to conjecture about a provable fact seems to be a terrible waste of resources. It is even more appalling that the pump manufacturer should have been forced to spend hundreds of thousands of dollars in its defense.

In what I thought was the end of the case, the jury took less than two hours to find in favor of the defendants. Two years later, attorneys for Gorman-Rupp contacted me again. The case had been appealed, and, based on procedural improprieties, a new trial was ordered. My testimony was required once again. A few more months passed before I learned that a settlement had been reached. I have no knowledge regarding the terms of the settlement.

I have testified as an expert in several cases since, but none has affected me quite as deeply as this one. The premise of the plaintiff's case was that a manufacturer had an obligation to make products that would not fail. In a little less than fifty years we have gone from a world where there was no ability to predict when something would break to a world where there is an expectation that all things should be built so that they will never break.

This expectation is unreasonable.

THINGS THAT DON'T BREAK

CHAPTER 6

There is just something that fascinates me about how things are purported to be unbreakable. Everything can be broken by the application of brute force. But what interests me is how frequently those things we think of as indestructible can be rendered impotent by the application of subtle influences. Two of my favorite "unbreakable" things are the Kryptonite brand of bicycle locks and so-called bulletproof glass.

When I was a student at MIT, all up and down Massachusetts Avenue there were Kryptonite locks apparently forever attached to parking meters, traffic signs, and bicycle racks, some of them still securing the remnants of the bicycles they were originally purchased to protect. It doesn't take much to imagine what had happened. Arriving for class, some student had used his "theft-proof" Kryptonite lock to make fast his bike, guaranteeing that it would still be there when needed for the journey home. At the end of the day, the student returns, reaches into his pocket, and, in stunned disbelief, discovers the key is missing. The less knowledgeable student would perhaps seek out a hacksaw or bolt-cutters, only to learn that his Kryptonite lock was, as advertised, unbreakable. Over days, parts of the bike would be scavenged—cables, brakes, derailleur—until all that remained was the lock, the bike frame, and sometimes a mangled wheel.

The flotsam and jetsam of bicycle locks and parts scattered up and down Mass. Ave. I took as a personal challenge—surely there was a way to break a Kryptonite lock. Not some obvious way, which anyone would think of, say using an oxyacetylene torch to melt through the lock, but an elegant approach.

I was not the only student thinking along the same lines. What distinguishes MIT students from their counterparts at other institutions is not technical ability. Many of the students at the Colorado School of Mines, where I now teach, are as technically capable as any I knew at MIT. What distinguishes MIT students is that they see even "little" problems as challenges in need of elegant solutions. These problems are so small, so irrelevant, that often to others they don't appear to be problems at all. So I was not surprised when one day the lunch conversation turned to breaking Kryptonite locks.

Two basic approaches were put forward. The first involved embrittling the locks. Mercury was an obvious embrittling element, but, being toxic and an environmental hazard, was not practical. Hydrogen was an alternative, less hazardous embrittling element. Many of those present at lunch had used electrochemical methods to charge steel specimens with hydrogen until they literally blew apart. But this approach required a power supply as well as flasks and beakers, and just didn't qualify as elegant. The second approach involved cooling the locks below their ductile-to-brittle transition temperature. While none of us knew exactly what the DTB temperature of a Kryptonite lock was, it certainly was above 70° Kelvin, the temperature of liquid nitrogen.

Liquid nitrogen is almost as common around a laboratory as water. So it was not long before we were all gathered around an abandoned Kryptonite lock, dousing it with the liquid nitrogen. As a thick layer of frost formed on the lock, we hit it with a stout hammer, and, *voilà,* it was broken. Within a few weeks the news of our accomplishment had spread, and abandoned Kryptonite locks passed into extinction in the vicinity of the MIT campus as hosts of students reproduced the now locally famous experiment.

Another "unbreakable" material that succumbs to subtle influences is Lexan, which is used as a replacement for glass when superior impact-resistance is needed. It is touted as a nearly unbreakable substitute for windowpanes and is frequently used when the possibility of vandalism is a consideration. It is used as a lens cover for automobile headlights because it is less likely to break when hit by flying road debris. On occasion, Lexan is even used as a security barrier, and is sometimes called "bulletproof glass," even though it is not a glass but a polymer.

Lexan is remarkable stuff. As is so often seen in movies and TV shows, it is probably a Lexan window that fails to break as the hero beats on it with a chair or other implement. However, Lexan is quickly embrittled by acetone—nail-polish remover. The hero would be better served by carrying a squirt gun filled with acetone instead of a .45. First, use the squirt gun on the window and then hit it with the chair. Actually, once it has been embrittled with acetone, even a light tap will break the formerly unbreakable window.

The previous discussion might encourage some bank robbers to change their MO, adding acetone to their toolkit. But this would be useless, since the partitions at most banks are made from real glass. All the acetone will permit the bank robber to do is remove fingerprints. The fact that glass can be made strong enough to turn a bullet may seem impossible. After all, isn't glass the most fragile of materials?

When I think of glass, one of the things that comes to mind is the sign that so frequently hangs in antique stores. "Lovely to look at, delightful to hold, drop it and break it, consider it sold." Oh, how I hated that sign. As a child I thought every china and glass store in the world had posted it just to induce a Pavlovian response in parents. They would see the sign, immediately turn to their children—particularly boys—and say, "Put your hands in your pockets and keep them there!" It didn't matter that I wouldn't even play with my marbles for fear of chipping them; my parents still felt some irresistible need to voice those words.

This conspiracy on the part of antique dealers, supported willingly or through conditioning by parents, serves to reinforce the illusion that china and glass are inherently fragile. Through most of my childhood, it was my belief that glass was an untrustworthy material; the slightest tap and a priceless object could be made worthless. It was for this reason that I took special interest in dishes manufactured by Corning and marketed under the name Corelle.

When introduced in the early 1970s, Corelle dishes were advertised as being able to stand up to the treatment handed out by a typical family, even one with boys. As I remember, commercials showed a boy engaged in some activity that would cause a stack of Corelle dishes to fall. The child would look as if the world had just come to an end as the dishes hit the floor with a clatter, but his parents would only smile, saying, "Don't worry, they're Corelle." As soon as I learned of these dishes, I had to have one.

Corelle dishes were not terribly expensive, but I couldn't buy just a single plate. At a minimum, they were sold as four-piece place settings. For my purposes, this was a bit more than I wished to spend, so I waited. Then one day, while wandering around the local discount store, there it was, an open four-piece place setting missing a piece or two. The remaining pieces were practically being given away. Needless to say, I snatched these up and hurried home with my find. The only question now was how to break my unbreakable dishes.

The dinner plate made an excellent Frisbee. Tentatively, at first, I simply threw the plate around the front yard. With each impact, I expected the plate to break, but it didn't. So I threw it a little harder and a little higher until I was heaving underhand and straight up as hard as I could. The plate would follow a nearly vertical trajectory, appear to pause at its apogee, and then plummet to the ground, often burying itself to half its diameter in the soft soil. But it did not break. Next I tried something a little more dramatic, skipping the plate off the sidewalk and road. Still the plate did not break, even when making five separate skips in one throw. Just as I was beginning to believe that Corelle was indeed indestructible, the plate made one final flight. This time it made a beautiful skip off the asphalt of the road in front of my house, arced left, and caught the concrete curb head-on. The plate didn't just break, it disintegrated into hundreds of pebble-sized pieces. This was one of the neatest things I had ever seen.

I thought I would be a little more scientific while breaking the teacup. I placed the cup upside down on the sidewalk and dropped my barbell weights on it from a stepladder. The teacup survived several glancing hits from a twenty-five-pound weight dropped from about eight feet. Finally a direct hit did to the cup what the curb had done to the plate, reducing it to hundreds of little pieces. Cool!

Corelle was just the latest in Corning's line of strengthened glass and ceramics products. The road to these "unbreakable" dishes began with a more serious problem:

Thousands of people were dying in train accidents. During the latter half of the nineteenth century railroad traffic was soaring and track congestion was becoming commonplace. The only effective means of regulating railroad traffic was a visual signaling system to tell the engineer what lay ahead on a given stretch of track. A kerosene-burning railway lantern with colored lenses served the purpose. The crucial components of these lanterns were the lenses, which were made from glass.

The word *glass* has two meanings. The first is general and refers to any substance with a particular arrangement of atoms. The second is more specific, referring to a particular substance with this structure. The glassy structure is one in which the constituent atoms are randomly situated, unlike crystalline materials, in which the atoms are arranged in a periodic array. As discussed previously, the atoms of glass are like the trees in a forest, whereas the atoms of crystal are like the trees in a nursery or Christmas-tree farm. The glassy structure is also called an amorphous structure.

A glass is formed when a molten material is cooled so rapidly that its atoms are trapped in nearly the same positions as in the liquid. These supercooled liquids, by definition, possess an amorphous or glassy structure. For most liquids, the cooling rate necessary to produce a glass is so high that these substances do not occur naturally. For others, however, such as common sand (silica), a very modest cooling rate will result in the amorphous structure. Obsidian is an example of a naturally forming silica glass. Other materials do form glasses. Many metallic alloys can be processed so as to produce a glass. These are collectively referred to as metallic glasses. Of all the glasses, the most ubiquitous is silica glass. Because of this, in the absence of any modifier, the word *glass* implies amorphous silica.

Among those properties that distinguish glass from other materials are its optical properties—glass can be made transparent to all or a portion of the visible spectrum, producing a glass of a given color. The craft of making and forming colored glass is ancient. Spectacular examples can be seen in the stained-glass windows of medieval cathedrals. Later the ability to produce sheets of uncolored glass revolutionized European architecture. It was the development of nineteenth century technologies, however, that ultimately forced the craft of glassmaking to become a science and to transform glass into one of the earliest "engineered materials."

Though the principal component of glass is silica, trace additions of other elements dramatically affect its properties—not only its optical properties but also its melting and working temperatures, its strength and resistance to chemical attack. Silica combines readily with a rich variety of chemical elements producing an almost infinite number of glasses with varying compositions and consequent properties. Early glassmakers held secret the compositions that they had stumbled upon. As technologies emerged that required materials with a specific combination of properties, the old craft of glassmaking inevitably gave way to science. Near the end of the nineteenth century, the German glassmaker Otto Schott undertook the first systematic study of the relationships between composition, processing, and properties. The young grandson of the founder of what was then Corning Glass Works, Alanson Houghton, was studying in Germany at the time and recognized the potential of a scientific approach to glassmaking. From that point on, R&D would be a supporting pillar at Corning Glass.

The first glass produced by Corning (c. 1890) was flint glass—composed from silica, lead oxide, and potash. Chemically stable, heavy, and expensive, it was used primarily in articles that required cutting or engraving. It was also used in products that were sufficiently crucial to justify the underlying cost and the expense of shipping, such as lightbulbs and railway lanterns.

The lantern lens of the time was formed from a series of concentric focusing bevels on the outside surface of the lens. These bevels had the unfortunate side effect of collecting snow, dirt, and other obstructions, reducing the lantern's range. To circumvent this problem, Charles Houghton (Alanson's uncle) designed a lens with the bevels on the inside. The placement of the focusing rings utilized the best optical knowledge of the day to increase the visibility of the lantern. With the advent of this lens and subsequent improvements, including formulations to produce standard lens colors, Corning all but monopolized the railway-lantern business.

Meanwhile, in Germany, Otto Schott had teamed with Ernst Abbe of Zeiss Scientific Instrument to monopolize another market for glass. They had developed a new family of tough glass with a number of attractive properties for laboratory glassware. By comparison with flint glass it expanded only slightly on heating, and thus was less susceptible to breakage from thermal shock. Additionally, it was remarkably resistant to laboratory chemicals. This combination of properties made a German company, Thuringian Glassworks, the undisputed leader in laboratory glassware, and there would be no serious attempt to challenge this market until World War I.

Researchers at Corning were intent on discovering the secret formulation of the German glass. And in time the secret found its way from Thuringian to the Reichanstalt (the German Bureau of Standards) and thence to the visiting scientist Arthur Day, who would return with it to Corning. The secret was borosilicate, used in place of the lead oxide of flint glass. These glasses are known as borosilicate glasses, which Corning patented under the trademark Nonex, or simply CNX (Corning Non-Expansion). From Nonex, Corning produced a "shatterproof" railway lantern globe. Unfortunately, Corning was a victim of its own technology, for while the product was clearly superior, railway lanterns had become virtually indestructible and the market for replacement lanterns flagged, necessitating the search for new consumer products.

Among the proposed uses for Nonex was as a line of housewares that would include pots and pans. As part of a well-publicized feasibility study, the wife of a senior Corning researcher, Jesse Littleton, baked a cake in a sawed-off battery jar made from Nonex. Her cake—or, more properly, the "pan" it was baked in—became famous, and transparent cookware was suddenly all the rage. Part of the reason for the sudden demand may have been related to the unsupported belief of the average consumer that transparent cookware was more sanitary than traditional metal pots and pans. Thus, Mrs. Littleton's latest culinary exploits, coupled with the perceived sanitary properties of glass, drove the campaign to incorporate Nonex into a line of cookware. Originally to be called "Py-right," reflecting the fact that the first commercial product was a pie plate, the product line was eventually trademarked under the name Pyrex.

Behind a successful marketing strategy, Pyrex cookware sold well from the outset. Sarah Tyson Rorer, editor of the *Ladies' Home Journal,* endorsed glass cookware for its efficiency. Jesse Littleton, whose wife had baked that first cake, conducted an ingenious demonstration to reinforce Rorer's point. He baked a cake in a pan that was half glass and half metal. The conventional wisdom was that the cake would not cook as rapidly in thick-walled glass as in metal, since the former was a poorer conductor of heat. However, the demonstration showed this not to be the case. In fact, the cake in the glass half of the pan cooked first. The explanation was simple: Heating resulted from radiant energy reflected from the oven walls. Unlike metals, which reflect this energy, glass is transparent, allowing the radiant energy to be absorbed by the cooking cake.

Transparent Pyrex cookware commanded the luxury price of two dollars for a large casserole in 1916. Corning was firmly ensconced in the housewares market.

By the late 1920s, the demand for Corning ovenware began to flag. To boost its consumer products division, Corning created a "test kitchen" run by Lucy Maltby, who had a strong background in chemistry and physics and a Ph.D. in home economics. The test kitchen was used to educate salespeople on cooking with Pyrex. During a week-long course, salesmen had an opportunity to "bake a cake, make tea or coffee or scalloped potatoes." But the test kitchen also provided photo ops; I have faint recollections of what must have been the earliest infomercials. These featured Lucy Maltby, who remained director of the test kitchen until 1965, and her staff educating the consumer on appropriate Pyrex cooking techniques.

From the outset, Maltby's goal was to make Pyrex into a basic cooking tool rather than the novelty item it was. She realized this goal with the introduction of a six-inch Pyrex skillet in 1932, though market research had shown that consumers most wanted a ten-inch skillet. However, developers considered this larger skillet unfeasible. The reason was simple: At the time, most range-top burners had a diameter of six or eight inches. A ten-inch skillet on one of these would experience differential heating and break from thermal stress. The managers, sticking by the research, were unwilling to relent until the laboratory personnel treated them to a cogent demonstration. Twenty-one ten-inch skillets were placed on twenty-one six-inch burners and simultaneously turned on. Within a few minutes three of the frying pans had broken in a more or less explosive manner. Management, recognizing the potential for catastrophe, authorized the six-inch skillet.

Cookware, of course, was not the only product produced by Corning. At about the same time the six-inch skillet was under development, Corning researchers were planning the fabrication of the 200-inch Pyrex mirror for the Mount Palomar Observatory. The demand for Pyrex laboratory ware was on the rise, and the new technologies were creating demands for glass with unusual properties. Television required glasses with high electrical resistivity; and architects were making ever more inventive use of glass. The Johnson's Wax Research Building, designed by Frank Lloyd Wright and built in 1952, was a showcase for glass in architecture. Its walls were formed from miles of Pyrex tubing.

It was against this backdrop that Corning took the next step toward tough glass for cookware, though this step was the result of a laboratory accident rather than a plan. Donald Stooky was heating a plate of glass to study its crystallization properties. The oven malfunctioned and instead of heating the sheet to 600° C. as planned, it was heated to 900° C. As Stooky recalls, "I expected to see a pool of melted glass, but to my surprise my plate still had sharp corners. I took it out of the hot furnace with tongs, but it fell to the floor with a clang like a sheet of plate steel, and it didn't break."

Though Stooky did not know it at the time, it would soon become clear that the increased temperature had caused regions of the glass to partially crystallize, as if transforming parts of a forest into orderly Christmas-tree farms. The resulting material is called a glass ceramic and, seemingly regardless of the composition of the glass, the partial crystallization has a strengthening effect. Corning developed the new glass under the name Pyroceram.

The potential properties of the new glass ceramic were ideal for Corning's line of cookware products, initiating a research program to introduce a new line of ovenware. Within months of Stooky's discovery, a low-expansion form of Pyroceram was perfected. From this new space-age material came Corning Ware—the flagship of Corning's consumer products division in the 1960s.

Even with the improved strength of Pyrex and Pyroceram, in the view of Bill Decker, the president of Corning, glass still had one serious drawback—it broke. It was this problem that Decker directed his researchers to fix, and with this directive, Project Muscle began.

Project Muscle began as an exploration of all known forms of glass strengthening. The most common approach to increasing the toughness of glass, dating back thousands of years, makes use of differential cooling to produce what is called tempered glass. Glass, like many materials, only breaks in tension. To illustrate this point, consider a simple example, familiar to many who have had the experience of cutting glass tubing.

When I was in junior high school, high school, and college, the first thing we did in chemistry lab was make a stirring rod. This was a pointless exercise, but it was fun. I don't think it is common practice anymore; at least it's not done at the university where I teach. The disappearance of the stirring rod is probably a consequence of litigaphobia. God help us if a student failed to follow directions and ran a glass tube through his or her hand.

Anyhow, the first step toward making a stirring rod is to cut about twelve inches of glass tubing from a much longer piece. To do this, one scores the tube gently with a small file. Then (this is the part that seems to confuse some students) grabbing the tube with both hands and placing thumbs opposite the scratch, the tube is bent away from the body. It does not take much pressure before the glass tubing has been divided into two pieces. Though these instructions may appear to be straightforward, every year somebody places his or her thumbs on the same side as the scratch. When these people bend the tubing, it does not always break transverse to the tube or in the intended place (opposite the thumbs), often resulting in badly lacerated hands.

What is going on here? When the glass is bent, the glass opposite the thumbs is in tension, while that on the same side is in compression. As the glass can only break in tension, the crack will begin on the side away from the thumbs. If there is an existing crack there—the one made by the file—the crack will run along this and exit the tube on the side near the thumbs. If there is no preexisting crack on the tension side, the crack will begin wherever the tension is greatest, which need not be immediately opposite the thumbs.

(A couple of interesting points, unrelated to the main discussion. First, if you really want to look as if you know what you are doing when cutting glass, spit on the scratch you've made with the file or glass cutter. Actually, any water source will do, but spit is handy. The hydroxide ions in water embrittle glass, making it easier to crack. Second, after cutting the tube, many minutes of fun can be had by polishing and bending the glass in a Bunsen burner. I am sure this is really the point of the lab—to play at shaping glass in a flame.)

The fact that glass breaks in tension can be used to advantage. If a plain piece of glass were placed in compression, to break it one would have to bend it more to overcome the compression. Picture it as a clothes pin with a rubber band wrapped around its jaws. By pushing the legs of the clothespin together, the jaws open up, stretching the rubber band—i.e., placing it in tension. Now wrap the rubber band around the jaws a few more times. The jaws are now in greater compression, and to get them open, a greater force must be applied to the legs. The same is true of the glass tube. If the surface away from the thumbs could be placed in compression, then this compression must be overcome before the tube will break. Fortunately, nature provides us with all the tools to put the surface of glass in compression.

Remember, most things, glass among them, contract on cooling. If the surfaces of a sheet of molten glass were suddenly cooled, they would shrink. The only thing resisting the shrinkage is the molten glass inside of the sheet, which, because it is liquid, doesn't provide much resistance; it just flows a little to allow for this contraction. Now we have two solid surfaces separated by a molten interior. As the interior cools, it also shrinks, but it is prevented from doing so by the already solidified surfaces. The shrinkage of the interior puts the surfaces in compression. In turn, the inside of the glass is in tension. The resultant material is called tempered glass.

In order to break tempered glass, the compressed surface must be placed in tension, requiring a greater force. Alternatively, a small crack in the surface, which penetrates the region of compression into the inner region of tension, would be unstable and grow spectacularly. This is why tempered glass, though tough, when it does break is often reduced to pea-sized shards all at once. The crack runs wild as soon as it enters the region of tension.

The amount of strengthening that can be realized by tempering is determined by how much the glass shrinks on cooling. Unfortunately, most glass compositions shrink only modestly on cooling. None of these would satisfy the goal of Project Muscle—an unbreakable glass. So the researchers at Corning sought a new approach to place the surface of glass under even greater compression. Their approach was ingenious—to use chemical methods to increase the distance between atoms in the glass surface.

The idea behind chemical strengthening is simple. Assume that we have a plain sheet of glass where there are no residual stresses—that is, both the interior and surfaces are under neither tension nor compression. Now, if we were to replace some of the atoms in the surface of this sheet with a larger type of atom, the surface would expand. However, the interior of the glass sheet would resist this expansion, because the atoms here would remain unchanged. So once again the interior of the glass would be under tension while the surface would be under compression, and the glass would be strengthened.

Changing atom types can increase their distances by 50 percent or more, whereas using differential thermal expansion and contraction generally increases the distance between atoms by less than 5 percent. As a consequence, chemical means offer a far greater strengthening potential.

In 1962, Corning held a press conference to make public its new line of chemically strengthened glass, Chemcor. There, sheets of glass were bent and twisted while others were tossed on the floor without breaking. The press was told of water tumblers made from Chemcor that had survived drops from the top of the nine-story research center at Corning.

Interestingly, Chemcor was not the kind of commercial success that Corning had anticipated. This was perhaps because Corning was unsuccessful in identifying product lines for their new chemically strengthened glass. Unlike previous products such as Pyrex and Pyroceram, which were immediately incorporated into new commercial products, Chemcor was a process waiting for an application. At its unveiling, Corning management opined, "[We] hope that not only design engineers but everyone throughout the country will be intrigued with glass so strong it can be bent and flexed repeatedly." Then they went on to add, "We feel other manufacturers may very well see uses for these new materials in other products—applications which we might not consider." This prompted one reporter's written comment, "Scientists have been dropping cups and saucers from the top of a nine-story building in Corning, New York, but don't ask them why. They're not sure."

Among the applications for which Corning attempted to develop a market for Chemcor were lenses for safety eyewear. Chemical strengthening allowed for both stronger and thinner, hence lighter, lenses—both attractive features in safety glasses. One manufacturer went so far as to bring Chemcor lenses to market, only to recall them when concerns were raised about the nature of the failure that might occur if the compression layer was scratched or pitted. Noting that because the surface compression in the chemically strengthened glass was so great that when it was scratched, "even a light impact will cause the lenses to 'explode'—

instead of losing an eye, the wearer could suffer brain damage."

The anticipated market for Chemcor never developed, though Corning continued research in chemical strengthening, perfecting twenty-three different chemically strengthened compositions by 1967. It was with these compositions and one additional strengthening approach that Corelle would be born.

The final hurdle in the development of "unbreakable" dishes came with the perfection of processing techniques allowing for the formation of glass laminates. Tempering produces compression in the surface by capitalizing on differential cooling in a single glass composition. Chemical strengthening similarly achieves surface compression through chemical modification in the surface layer of a single glass composition. With a single glass composition, the extent to which the stress distribution can be controlled is limited. However, if one could form a laminate from glasses of several different compositions, each with different thermal (expansion) and chemical properties, much greater control of this stress distribution would be permitted. For instance, consider a laminate made from five layers of glass. In principle the two surface layers and the middle layer could all be placed in compression, each separated by a tension layer. Now a crack penetrating the surface layer would run through the tension layer, only to be stopped by the middle compression layer.

As is so often the case in the world of materials, it is not the conceptual design that poses problems, but actualizing this design. How does one form glass laminates? This is a little like trying to make ice cubes with alternating red and green layers for a Christmas punch; try this, it is doable. Given the time and effort invested, however, you will not enjoy watching your ice cubes melt away. Unless, of course, you have adequately partaken of the punch before adding the ice cubes. Even so, after your head has cleared, it is unlikely that you will start a successful business selling laminated ice cubes. That is, unless you can come up with some ingenious technique to make the whole process financially reasonable. This is what Jim Giffen did for Corning. He perfected the processes for making glass laminates.

Capitalizing on Corning's historical understanding of the marketplace for housewares, using Giffen's techniques to form glass laminates from Chemcor compositions, Corelle Livingware made its debut in 1970. By 1977 Corelle had snagged an astonishing 20-percent market share. And on February 29, 1984, Corning's steam whistle blew to signify the production of the billionth piece of Corelle. All I know for sure is that at least one dinner plate and one teacup have been broken, but it seems clear that the vast majority of those billion dishes are still intact. And on occasion while wandering through an antique store I see a piece of Corelle and look around for that sign, "Lovely to look at, delightful to hold, drop it and break it, consider it sold," and I chuckle while thinking, "Drop it and break it—not likely."

WHEN THE GOING GETS TOUGH

CHAPTER

7

Many materials can be made strong, as in the case of Corelle Livingware. Strength tells us something about how much load something can carry—an important consideration for many applications. Steve Trotter may have learned this as he and four friends prepared to swing from the Tampa Bay Skyway Bridge on April 27, 1997. The idea was simple: Attach one end of a 180-foot cable to the bridge's high point, about 200 feet above Tampa Bay; next, fully extend the cable along the bridge until taut, then jump over the side. If everything went as planned, the five thrill-seekers would have one hell of a ride—unlike anything you find at a municipal playground—accelerating to over 85 mph in just a few seconds.

This would not be the first time Steve had made himself into the weighted end of a giant pendulum. Reputedly, he had invented the stunt, previously swinging from the Golden Gate Bridge, the George Washington Bridge, and others. Unfortunately, the requirements for a pendulum swing launching pad are fairly restrictive. The launch site must be nearly twice as wide as high. And bridges are about the only structures that satisfy this requirement. So, after Steve had jumped from all of the really big bridges, in order to keep "pushing the envelope" he had to come up with something new. His solution—add passengers to the swing.

In 1990, Steve, Justin Porter, and Glen Rohm had swung together from the Tampa Bay Bridge. Seven years later, five swinging together from the same bridge would set a new "world's record," at least according to Steve and his fellow would-be record-setters. So, alerted to Trotter's intentions, a helicopter with a video crew was on hand as Steve and his companions, Kenny Bunker, Glen Rohm, Lorie Martin, and Jeff Sergeant, piled out of a car to which the steel cable had been tethered and began to run along the bridge. The five had to hurry. Police do not look favorably on such stunts, and would undoubtedly arrest the group if they found them linked together and attached to a steel cable fixed at one end to a parked car. Their intentions would not be hard to guess. So, only a few seconds after parking the car, the thrill-seekers leaped off the bridge.

About sixty feet above the water, the steel cable broke, and the swingers became fallers, slamming into the water of Tampa Bay at over 70 mph. As recorded by the video crew, Lorie and Jeff were knocked unconscious and floated face-down as Kenny, the only uninjured member of the team, struggled to revive them. Fortunately a boat came to the rescue, removing all five from the water before anyone drowned. In all, the members of the team suffered injuries ranging from broken necks and ribs to punctured lungs.

As you might suspect, with the dramatic video, the tale of the "stunt gone wrong" became the subject of television news magazines, including NBC's *Dateline.* It was while watching this show that the full circumstances surrounding the accident became clear to me. Apparently Trotter had given some thought to the strength of cable upon which he and his team members would depend. He chose a cable with a tensile strength greater than the combined weight of the five swingers, about 900 pounds. In fact the cable—with a diameter about that of a man's pinky finger—had a tensile strength of about 1,800 pounds. As far as Trotter was concerned, this should have been more than sufficient. Sadly, Trotter knew little about basic physics.

I'm sure when Steve was young he had occasion to swing a weight tied to a string about his head, as most of us have. And he must have felt the centrifugal force pulling the string tight. This same force was at work as the five adventurers swung above Tampa Bay. Simple calculations show that the centrifugal force, when combined with the weight of jumpers, would yield a tension in the cable of about 2,700 pounds—a tension that would not be realized as the strength of the cable was only 67 percent of the minimum required.

Surprisingly, when interviewed for *Dateline,* Trotter seemed completely unaware and remarkably unconcerned about his "design error," and boasted that he would try the stunt again. Before he does, I urge him to learn a little physics and a little about the strength of materials.

Strength is just one of a number of properties that determine when something will break. When considering fracture, another equally important materials property is "toughness." When used to describe a person, this word conjures up the image of a rugged, strong, and independent individual, which is true, but it lacks the precision needed for a scientific definition. In science a property or characteristic is most useful when it can be measured. So the question of defining toughness is reduced to specifying the manner by which toughness is to be measured.

Before delving into the measurement of toughness, we need to review a couple of nature's laws. One of these is that energy is conserved—the so-called First Law of Thermodynamics. For those in the know, it is often referred to by its nickname, the First Law, taking the same place in the scientist's pantheon of laws as "I am the Lord thy God" does in the Ten Commandments of Moses. The First Law merits special attention because it is more than a conservation law. Not only is energy conserved, the First Law also tells us that energy can only take two forms—heat and work. Throughout the universe, all change is a manifestation of nothing more than transforming a fixed amount of energy between its two forms. All change will cease when it is no longer possible to transform heat into work and vice versa. But that has to do with the Second Law, which we will get to in due course.

Just the fact that energy is conserved, not that it has two forms, tells us something important about the history of the universe. If you were to journey back to any time in the past and measure the total amount of energy in the universe, you would get some number; call it 1 Eu for "energy of the universe." Now if you were to repeat the experiment, only this time going forward to any time, you would get the same result—1 Eu. The energy might have changed form between your measurements. In the past it might have been found in the work needed to separate interacting particles, while in the future that same energy is found in the heat released as these particles combine into nuclei, atoms, and gravitating planets. But regardless of the form the energy takes, its total amount remains unchanged as long as the universe exists.

Though almost every educated person knows that energy, among other things, is conserved, the lesser-known implications of conservation laws are subtle and profound. It turns out that because energy is conserved, we know that the time at which an experiment is done does not affect the outcome of that experiment. Now, I can imagine you saying, "That's absurd. If I measure the temperature, I clearly get a different result depending on the time of the measurement. It is cooler at night than during the day, and in the winter than in the summer." All of this is true, but those differences are the result of changing influences—the amount of sunlight hitting the air and stirring up the molecules changed between night and day and summer and winter. If all those things that can influence temperature were held fixed and only time was allowed to move forward, you would always get the same temperature—time alone has no effect on your measurements.

Because "when" an experiment is done is irrelevant to the outcome of that experiment, we know that the natural laws governing the universe ten billion years ago are no different from those at work today or those that will be at work ten billion years in the future. We know all of this because energy is conserved.

Energy is not the only conserved quantity—angular momentum is another. Conservation of angular momentum is what guarantees that the Earth will continue spinning on its axis, the moon revolving around the Earth, and the Earth about the sun. Though all of these are slowing down, this is due to collisions with particles floating about in space, not because of some mysterious loss of angular momentum. As the moon, for example, revolves about the Earth, it collides with space dust, transferring a little angular momentum with each collision. The total angular momentum of the moon and the space dust remains unchanged, but now the moon has a little less and the dust a little more. Over a great deal of time, the moon will lose its angular momentum to space debris. In a simple system, where a single light object orbits a much more massive planet or star, as the object loses angular momentum its orbit shrinks. Eventually it will collide with the surface of the planet or star. But don't worry; the moon is not about to collide with the Earth. For though it is losing angular momentum to space dust, the moon is also gaining it through the slowing of the Earth's rotation about its axis. As a consequence, the moon is receding, while the length of a day is increasing. If you can't get everything you need done in a twenty-four-hour day, just wait a few million years.

Like the conservation of energy, conservation of angular momentum has subtle implications as to the outcome of experiments. With angular momentum conserved, we know that the outcome of an experiment will be the same regardless of the direction you are facing when you perform that experiment. In other words, nature's laws are isotropic—there are no special directions to the universe as there are in, for example, crystals.

Another conserved quantity is linear momentum, often simply referred to as momentum. Momentum is determined by taking the product of mass with velocity. For example, a 100-kilogram mass (220 pounds) moving at three meters a second (six miles per hour) would have 300 units of momentum. (A common unit of momentum is called a Newton second, and is expressed as N sec = Kg m/sec). The conservation of momentum is what you experience when someone runs into you. The momentum lost by the person running into you, you must gain. So if a person were to lose 100 units of momentum as they collided with you, you would have to *gain* this same amount. Assuming you weigh 50 kg, your velocity would increase by two meters per second (about four miles per hour) as a result of the collision.

The conservation of linear momentum implies that your position in space makes absolutely no difference to the outcome of an experiment. Whether the experiment is performed here or in the Andromeda galaxy, the results will be the same, if all other influences are held constant. The laws of nature do not change from place to place, direction to direction, or time to time. We know this because linear momentum, angular momentum, and energy are all conserved.

In the discussion of conserved quantities, I saved linear momentum for the last because the word is so often misused. I find the misuse of scientific words particularly irritating, as it tends to erect further barriers between the public and the concepts of science and technology. Steve Trotter's little mistake, which could have resulted in five deaths, serves to illustrate the possible consequences of those barriers. It is difficult to understand how a society increasingly reliant on technology can continue to function as our collective understanding of the science that drives this technology seems to diminish at a nearly equal rate!

Returning to the misuse of the word *momentum*: Momentum is conserved except on weekends and some holidays between mid-August and January, when it becomes the most fickle of quantities. It is during this period that sports commentators choose to explain sudden shifts in the fortunes of one football team or another as a loss or gain of momentum: "Well, Bud, with that interception the Broncos have lost their momentum." Two plays later the Broncos may well "regain their momentum." Because momentum is conserved, these comments naturally invite the questions "Where did the momentum go?" and "From where did it come?"

If sports commentators are in need of a conserved quantity to describe football, then I suggest they use net yardage.

For every yard gained by one team, the opposing team loses a yard. Taking yards gained as positive and yards lost as negative, the sum must always be zero. Hence, net yards are conserved.

Now the commentator can say, "Bud, with that twenty-yard gain by the Bronco's running back, the Falcons have lost twenty yards." It sounds inane, but hey, these guys may only get a million dollars a year for adding "color" to football games. (My favorite inane football commentator quotation followed a pair of touchdowns by the Denver Broncos in answer to a single scoring drive by the opposing team. The color guy, feeling the need to say something, came up with a real gem: "You don't want to get into a scoring contest with the Denver Broncos." Excuse me, isn't that what we're watching here—a scoring contest?)

After this somewhat lengthy discussion of conserved quantities and the correct usage of scientific terms, let's return to the measurement of toughness. Picture a sledgehammer suspended by a pin through its handle, such that it can swing freely about this pin. We will call the position of the hammer when hanging straight up and down its *equilibrium position*. When the hammer is displaced to the right or left from this position, the head of the hammer is elevated above its equilibrium position. Upon release, it oscillates back and forth like the pendulum of a grandfather clock. If the apparatus is well made, with little friction between the pin and the handle, every oscillation will be of the same amplitude—the head of the hammer will always rise to the same height above the equilibrium position. This is because energy is conserved.

Energy has been "stored" in the hammer as work by displacing it from its equilibrium position. The amount of stored energy is proportional to the change in elevation of the hammer. When dropped, this stored energy is released as motion. At the bottom of its swing, all of the stored energy has been converted to energy of motion. At the top of the swing, this energy is once again stored in the hammer. Because energy is conserved, the hammer returns to the same height on the right as the original elevation on the left. If it did not, some energy would have disappeared. Of course, in a real experiment the elevation gain on the right is just a little bit less than the original elevation. This difference in elevation is proportional to the energy lost by the hammer to molecules in the air. A very accurate thermometer would show that the temperature of the air increased (work converted to heat) as the oscillating hammer "wound down." For our purposes, however, we can ignore that tiny amount of energy transferred to the air during a single oscillation.

Now, add to the apparatus a clamp, which is firmly fixed to the ground and positioned just below the head of the hammer in its equilibrium position. If there is nothing in the clamp, the hammer will just miss the clamp on each swing. Now—and this is the fun part—let's put something in the clamp, say a Corelle dinner plate. Displace the hammer to the left. Carefully record its elevation and let go. The hammer swings, hits the plate, and shatters it into thousands of pieces. But don't stop to look at the remnants of the plate just yet. Watch the hammer until it reaches the top of its swing on the right and record its elevation. In this case, the hammer does not swing as high on the right as when it started on the left. Because energy is conserved, the energy stored in the hammer did not disappear but was lost to something else, but what?

Well, some of it was changed to sound energy—we heard the fracturing plate. But this represents a very small fraction of the missing energy. Another very small fraction was converted to thermal energy—the pieces of dinner plate will now be a little hotter than before. But, yet again, this represents only an insignificant part of the missing energy. The bulk of the energy went into breaking the chemical bonds that had formerly held the plate together. This is the work of fracture or fracture energy, and is related to the material's toughness. The more energy it takes to break something, the tougher it is.

Let's repeat the experiment, only this time replace the dinner plate with a piece of copper of equivalent cross-section. Lift the hammer to the same height as in the previous experiment and let go. Again the hammer will hit the specimen and break it. This time, however, there will be only two pieces of copper, as opposed to the thousands of pieces of broken Corelle. Additionally, the energy lost by the hammer in breaking the copper will be much larger than that lost in breaking the dinner plate. How can this be? There was so much more "fracture surface" when breaking the strengthened glass, many more bonds must have been broken. From this one might mistakenly conclude that strong Corelle Livingware has a higher toughness than copper. Not so. So, when the copper was broken, where did the additional energy go?

Careful examination of the broken copper specimen supplies the answer. Near the fracture surface there are slip bands. The energy lost by the hammer went into making, moving, and piling up dislocations, that is, into the work of bending the copper before it broke. Tough materials are those possessing a channel, other than fracture, in which to store energy. In the case of copper and many other metals, this additional channel involves making and moving dislocations. This is why metals are usually tougher than glasses or ceramics.

In general, most structural metals and alloys are 50 to 100 times tougher than glasses and ceramics. Surprisingly, there is an inverse correlation between strength and toughness. Mild steels may be as much as three times tougher than so-called high-strength steels, and a high-strength aluminum alloy only half as tough as its not-so-strong counterpart.

While most metals owe their toughness to the making and moving of dislocations, and the toughness of ceramics is the result of making surface, there are other classes of tough materials where dislocations and making surfaces play little or no part. A group of polymers known as aramids form one such class. DuPont markets the best-known, and probably first, aramid under the name Kevlar.

Kevlar is the material from which I was building kayaks in the early 1970s. And it was observing how these boats broke that motivated my decision to go to graduate school. For this reason I have a sentimental attachment to Kevlar, having watched it grow from a little-known product to a superstar in the polymer world.

When the company first introduced it in 1971, DuPont

was promoting this new material in a number of products including sporting goods. At that time only a handful of people were building Kevlar kayaks, and most of them were located in Colorado. We had all chipped in and placed what seemed to us a very large order for the new wonder fabric. As a consequence, I was not at all surprised when representatives from DuPont contacted me to say they would be passing through Denver and would like to meet with some of the boat builders in Colorado. Several of us took the guys from DuPont up on their offer, joining them for dinner at a local restaurant

I was amazed to learn that our Kevlar order was almost insignificant compared to other applications. One of these was for boat sails. Light and strong, Kevlar is an ideal material from which to make sails. A single set of sails for a twelve-meter yacht would contain more Kevlar than a hundred or more kayaks. Even more surprising was the revelation that foreign governments were trying to buy Kevlar. Evidently it was so tough it could be used as a material for bulletproof vests. This application posed a dilemma. DuPont did not want its product used by a hostile government in a way that would jeopardize the interests of the United States, but still wished to market Kevlar widely for peaceful purposes. How DuPont resolved the dilemma, I don't know. However, some thirty years later bulletproof garments made of Kevlar have saved the lives of hundreds of law-enforcement agents and are marketed freely over the Internet.

Kevlar is a linear molecule that interacts with other molecules to form rod-shaped structures. Think of a single molecule of Kevlar as a Popsicle stick, and the rod-shaped structures as bundles of these held together by rubber bands. If you hold the bundle in your left hand, you can grasp the end of a single stick with your right hand and rotate it around its long axis. As you do so, the rubber band may expand a little to accommodate the energy you are imparting to the bundle. If you impart a large amount of energy to the bundle, say by hitting it with your fist, none of the Popsicle sticks will break. Instead the bundle will simply flatten out as each stick rotates, stretching the rubber band and absorbing the energy from your fist. This is the mechanism from which Kevlar derives its toughness. Of course, the molecules of Kevlar are not held together by rubber bands, but by intermolecular bonds.

Tough and strong as Kevlar is, it is not the toughest polymer known. That distinction goes to spider silk. Think about it. A strand of spiderweb has a diameter only a small fraction of that of a human hair. Yet it can absorb the impact of a large moving insect without breaking. It has been said that a half-inch-diameter rope made of spider silk could stop a Boeing 747 in flight.

Though we know the sequence of elements that makes spider silk, how the spider forces the molecules of this polymer to arrange the right way is still a mystery. Materials specialists at universities throughout the world have been working on the problem of synthesizing spider silk for years, but for now, man-made polymers are a poor substitute for the truly tough materials nature has produced.

There are also novel mechanisms used to toughen ceramics. My favorite product made from these materials is a common hammer. A hammer made from steel hardly deserves a second look. But one made from a ceramic—that is intriguing.

I saw a ceramic hammer while touring the Coors Ceramics facility near the School of Mines in Golden, Colorado. Though Coors is well known as a brewer of beer, it is less well known as the world's largest producer of specialty ceramics products. This particular hammer was made from transformation-toughened zirconia, a polycrystalline material made from the oxide of zirconium, a metal. Though a ceramic hammer may not be a practical replacement for the more common metal variety, the fact that one can be produced at all provides a powerful visual image for those considering Coors as a manufacturer of their product.

Ceramics offer a number of advantages over metals. Most important of these is that they do not burn. Most metals burn to form the oxide of that metal. Often this does not pose a problem, because the temperature necessary to start the metal burning is so high that for all practical purposes the metal can be considered inert. For example, an aluminum can does not suddenly burst into flame, but under the right conditions it will. The British Navy learned this lesson in a tragic way during the Falklands crisis in 1982.

For cost and performance reasons, many British warships are built of aluminum. Though light, aluminum has a serious disadvantage compared to steel: It burns spontaneously at much lower temperatures than iron. This point was driven home rather dramatically during the Falklands War, when Argentine Exocet missiles and bombs sank six British ships including HMS *Sheffield, Antelope, Ardent,* and

Coventry. Some have speculated that the explosions by themselves should not have caused these ships to sink. However, the subsequent fires were so hot that the superstructure of these ships literally caught fire and burned. If the ships had been made from steel, this would not have happened, for while steel burns to form an oxide, the temperature necessary to start iron burning is extraordinarily high.

Though metals are useful, there is always this problem lurking out there: How do we stop them from burning? The problem becomes particularly acute in a high-temperature environment such as a rocket nozzle or a jet turbine. It just would not do if these structures were to catch fire and burn. On the other hand, a ceramic does not burn. It is produced from the burning of a metal. Zirconia is simply the ash left after zirconium has burned. As a result, ceramics are the ideal materials to use for high-temperature applications. If for some reason you needed to drive nails in a really hot environment, you would need a ceramic hammer. Unfortunately, ceramics are not very tough. So, though there is no risk of a ceramic jet turbine burning, if the same turbine were to suck in a bird, it would shatter.

Crashing, whether due to a burning or a shattering turbine doesn't provide the turbine designer with acceptable choices. It is for these reasons that much high-temperature materials research is directed toward two ends: first, making metals that won't burn, and, second, making ceramics tougher. It is the second research objective that makes ceramic hammers so interesting.

As the name implies, transformation toughening relies on a transformation between two or more solid phases.

Though we are familiar with the fact that most substances can exist in one of three forms—solid, liquid, and gas—there is less recognition that many substances can exist in different solid forms. Carbon is a well-known material with several solid forms. Most familiar are graphite and diamond. Though both are made exclusively of carbon atoms, these atoms are packed together in different repetitive patterns to give materials with different properties. Graphite is the stuff that forms the working part of a pencil. It is black and soft. Diamonds are an essential component of Oscar night. They are hard and transparent.

Zirconia has three different solid phases. Perhaps the most familiar of these is cubic zirconia, the diamond substitute in costume jewelry. As the name implies, the atoms of zirconium and oxygen are packed together in a cubic array. For the other two solid phases, the constituent atoms are packed together in a tetragonal or monoclinic array. Unlike diamond and graphite, transformations between these three phases of zirconia are triggered by changes in temperature. From room temperature to about 950° C., the monoclinic structure is observed. But increase the temperature above 950° C., and monoclinic zirconia transforms to the tetragonal phase. At still higher temperatures there is a tetragonal-to-cubic transformation.

I'm sure some readers note an inconsistency here. If cubic zirconia is only stable at high temperature, how can it be used in costume jewelry? I have yet to find heated jewelry display cases at local department stores. Sometimes trace amounts of impurities have a dramatic effect on the stability of different solid phases. In the case of zirconia, a little bit of calcium will make the cubic phase more stable than either the tetragonal or monoclinic phases at room temperature.

All of this aside, it is the transformation between tetragonal and monoclinic phases that is used to toughen zirconia. The tetragonal phase is denser than the monoclinic phase. So, on cooling, the phase transformation at 950° C. is accompanied by an expansion. With appropriate impurities and cooling rates, zirconia can be made so that small, lens-shaped precipitates of the tetragonal phase are retained in the cubic phase all the way to room temperature. These precipitates cannot change phase because the surrounding stable phase prevents the expansion necessary for the transformation.

Consider what would happen to a crack running through this material. When the crack passes through the tetragonal phase, it transforms, as it is no longer constrained to a volume too small for the monoclinic phase. Accompanying this transformation is an expansion that literally clamps down on the tip of the propagating crack, stopping it in the midst of its destructive growth. To fracture such a material requires energy to push the crack through the closing jaws of the transforming tetragonal phase. Hence the material is toughened through these phase transformations. Neat.

Tough as transformation-toughened zirconia is, it is still only one-fifth to one-tenth as tough as most common metals. However, transformation toughening is not restricted to ceramics or to zirconia. A similar mechanism is used to toughen some forms of steel—making for some *very* tough stuff.

Among the applications that most require tough materials is the production of armor. Typically, one thinks of armor as made of metal, but there are many instances where the weight of metal armor makes it impractical. In these instances, ceramics have become the materials of choice.

For aircraft, weight is of paramount concern. From the outset, advances in aviation have been reliant on lightweight materials. At first, planes were built of wood, paper, and canvas. These were fine materials for planes that flew only slightly faster than a man could run, but as plane performance improved, plywood was substituted for paper and canvas. Perhaps surprisingly, some combat aircraft of World War II were built from wood. The Royal Air Force's Hawker Hurricane and the DeHaviland Mosquito were of all wood construction, affording the latter the nickname "the Wooden Wonder." Its construction did nothing to hamper its performance; so formidable was this fighter/bomber that the Luftwaffe awarded two "victories" for each one destroyed.

Wood is an amazingly versatile structural material—light, strong, tough, and easy to fashion into a multitude of shapes. However, wood has one ugly property that limits its role in combat aircraft. It burns! And it doesn't take much to get it going. As a result, wooden aircraft are particularly susceptible to incendiary bullets. A bullet strike anywhere on a wooden plane can result in fire and the subsequent loss of the aircraft. For metal aircraft, only a hit to the fuel tank is likely to cause a fire and an associated crash.

At the same time the vulnerabilities of wood aircraft were exploited, the operating temperatures of aircraft were climbing. With the advent of jet aircraft, not only were the materials used in the engines exposed to unprecedented temperatures, but so also were many elements of the aircraft's superstructure. In supersonic jets, air friction on the leading edges of wings often raises them to temperatures well above the combustion point of wood. What made the development of jets possible was the availability of new metallic materials that could perform at higher temperatures. New nickel alloys replaced iron-based materials in aircraft engines, and aluminum and occasionally titanium alloys replaced wood as the primary building material of aircraft superstructure.

While the aluminum and titanium skins of modern aircraft may be tougher than plywood, none of these materials is intended to resist antiaircraft fire. This does not pose a great risk, for a bullet randomly hitting a combat aircraft will simply pass through, leaving entry and exit holes that seldom compromise the planes' integrity. There are, however, places where a bullet strike may be catastrophic; a hit to the fuel tank, hydraulics, or cockpit are examples. These regions are often armored, and because weight is still a factor, ceramic armor is the choice.

Ideally, ceramic armor would be just as tough as metal—absorbing the energy of the projectile. Yet, even with transformation toughening, ceramics offer only a fraction of this energy dissipation. But there is another property that is advantageous for armor—hardness.

Hardness characterizes a material's resistance to penetration, and has been used to determine quickly whether a substance is what it is claimed to be. Real diamonds, the hardest of materials, can be distinguished from imitations if they scratch glass. Though I have never tried, it is reputed that the purity of a gold coin can be ascertained by biting it. Pure gold is softer than its alloys or other materials that are golden in color. As such, pure gold will show bite impressions where imitation gold will not. A measurement of hardness is provided from the depth to which a given force acting on a standard shape penetrates an object. One of the more common of these measurements is called the Rockwell C hardness test. The indentation caused by a diamond cone supporting a 150-kilogram weight is determined, and this depth is then converted to a "relative hardness." On this scale the hardest substance known, diamond, has a Rockwell C hardness of 100, whereas an aluminum can has a hardness of 20.

Typically, hardness and toughness are mutually exclusive properties. Very hard materials are seldom if ever tough, and vice versa (just as strength and toughness are mutually exclusive). So what is the advantage of hard armor? The whole idea of armor is to dissipate the energy of a projectile before it makes contact with a human being or critical components. However, as long as it is not a person or a critical component, it doesn't matter what dissipates the energy. When a projectile—usually made of tough but not particularly hard metal—hits a hard surface, it is the projectile that deforms, dissipating its energy of motion. The theory behind hard armor is for the armor to do damage to the bullets.

This is exactly how the ceramic armor protecting the cockpit of the U.S. Air Force's C-141 works. The interior of the cockpit is covered with tiles four inches square. Each tile has two layers. The first consists of tough steel, about .03 of an inch thick; the second, about three-quarters of an inch thick, is made from a hard ceramic, alumina (what one gets after burning aluminum). When a bullet hits one of these tiles, it is first deformed by the hard ceramic. If there is sufficient energy remaining, the bullet breaks the ceramic, further dissipating its energy. Yet more energy is removed from the bullet as it is eroded by its transits of the now broken but still hard and abrasive ceramic. What energy is left in the bullet is then expended deforming the tough metal layer.

Ceramic armor tiles are expected to break. In part, the protection offered by these tiles requires them to break. Imagine the crew of a C-141 returning from a mission after having taken small-arms fire. Everyone is alive and well, thanks to the armored cockpit, but many of the tiles are broken. Obviously those life-saving tiles should be replaced before the next mission. No problem. The tiles are attached with Velcro to the inside of the cockpit. They're just peeled off and new ones are pressed on.

It may seem improbable that a tile attached by Velcro would not be dislodged by a bullet, but this is the case. Sometime try pulling a Velcro attachment apart without first separating it at an edge. It can't be done. In fact, the Velcro itself offers another toughening mechanism for the ceramic armor of the C-141 aircraft.

With every advance in armor comes an advance in antiarmor. Projectiles sporting shaped charges will penetrate virtually anything, regardless of how tough or hard. A shaped charge uses the energy of a high explosive to form a very fast-moving jet of molten metal. It is much like squeezing a tube of toothpaste. In this case, copper takes the place of toothpaste and explosives take the place of the squeeze. The armor-piercing projectile, or slug, is a high explosive with a copper-lined conical cavity. Upon impact the slug detonates, forcing the copper into a conical jet moving at over 30,000 feet per second. That's thirty times the speed of sound, 20,000 miles per hour, ten times the speed of a high-velocity bullet. The resulting jet is a formidable weapon, eroding anything in its path. A small shaped charge, about two inches in diameter and four inches long, will penetrate ten to twelve inches of armor steel. The larger shaped charges used in the U.S. TOW antitank missile will penetrate twenty inches of steel. Because the copper jet owes its penetrating power to erosion, it makes very little difference whether the armor is ceramic or metallic. In either case, the armor is defeated.

With every advance in anti-armor comes an advance in armor. To counter shaped charges comes reactive armor. In its simplest form, reactive armor is formed from two plates of armor sandwiching a layer of high explosive. This sandwich "reacts" to the jet's impact, detonating its explosive meat. The explosion propels the armor outward, deflecting the erosive jet. More-advanced applications of reactive armor yield a thirty-five-fold improvement over steel armor.

With every advance in armor comes an advance in anti-armor. Countering the shaped charge with reactive armor exposes a serious vulnerability: It leaves behind a hole in the first layer of armor as well as in the explosive layer underneath. A second projectile hitting this spot could more easily penetrate the remaining layer of armor. But what is the probability that a second round will hit exactly the same spot? High, if the armor-piercing round consists of two slugs separated by only a few feet. The first round detonates the reactive armor, creating the vulnerability, and a fraction of a millisecond later comes the second slug to exploit this weakness.

With every advance in anti-armor comes an advance in armor. Stay tuned.

Only the Tough Get to Go

CHAPTER 8

Early Wednesday morning, October 15, 1997, a three-judge panel of the Ninth U.S. Circuit Court of Appeals refused to issue an injunction to stop the launch of a Titan 4B booster. Hours later the Titan lit the skies over Cape Canaveral. Atop the rocket sat the *Cassini* unmanned probe to Saturn, scheduled to reach the ringed planet in June 2004.

To gain the momentum necessary for this journey, *Cassini* would slingshot twice around Venus, once around the Earth, and finally around Jupiter, each time imperceptibly slowing the pivot planet while it accelerated toward its final rendezvous. Once there, *Cassini* will spend a minimum of four years observing Saturn and its rings and moons, including the giant moon Titan, and transmit the results of these observations back to the waiting earthbound scientists. The energy needed to gather this information and relay it home is generated by the radioactive decay of seventy-two pounds of plutonium.

Plutonium. The word evokes images of mushroom clouds, radiation sickness, and nuclear devastation. So it was not surprising that some reacted with alarm on learning that this most terrifying of substances was to be launched into the Florida sky. What would happen if *Cassini* were to be destroyed during launch by a malfunctioning booster, much the way *Challenger* had; or, worse yet, during its Earth fly-by, when the probe could deviate from its intended trajectory and plunge into the atmosphere? Would the plutonium be vaporized, possibly raining death and destruction across the entire planet? Buttressed by these concerns, several groups, including Hawaii's Green Party and the Florida Coalition for Peace and Justice, sought an injunction to prevent the launch. While the injunction request worked its way through the legal system, others expressed their concerns more visibly. The week before the launch, a thousand anti-*Cassini* protesters rallied outside the gates of Cape Canaveral Air Station. Though most of the protesters were well behaved, twenty-five were detained and arrested as they climbed a fence in an attempt to gain access to the military facility. Clearly, those opposed to the *Cassini* launch were genuinely concerned about the risks posed by the plutonium power source—risks that NASA and the Department of Energy had considered and spent over forty years addressing.

There are three alternative ways to power space probes: solar voltaics, batteries and fuel cells, or radioisotope thermoelectric generators (RTGs). The first of these, solar voltaics, "harvests" the energy of photons, converting it to electrical power. This harvesting takes place by intercepting solar radiation with large "photon nets" called solar panels or arrays. The power generated using these devices depends on the number of intercepted photons, which, in turn, depends on the size of the panel and the intensity of the solar radiation. Hence, as one moves farther away from the sun, where its intensity decreases, larger panels are required to generate equivalent electrical power. To power the *Cassini* probe, which will be operating almost a billion miles from the sun, would require solar panels the size of a football field.

The machinery necessary to keep such large arrays pointed at the sun is complex. Also, an array of this size would increase *Cassini*'s moment of inertia, making it more difficult to turn and maneuver. Adding additional complexity to a satellite increases the probability that something will go wrong, and any problems with the probe's power supply could lead to complete mission failure. So solar voltaics were considered impractical for the $2-billion *Cassini* mission.

Batteries and fuel cells, the second alternative, convert chemical energy to electrical energy. Unlike solar panels, which use energy from the environment, a battery's power comes from chemical fuel contained within it. The problem is that fuel adds weight. Generating the necessary energy to power *Cassini*'s four-year mission would require that some instruments be sacrificed to compensate for this weight. This seemed an unnecessary compromise in light of the third alternative, RTGs.

RTGs were developed by the Atomic Energy Commission in the latter half of the 1950s to provide electric power during space missions. Originally dubbed "atomic batteries," RTGs convert the heat generated by radioactive decay into electrical current. Like batteries, the "fuel" to power an RTG must be supplied. Unlike batteries, the weight of this fuel is not excessive because much more energy is liberated from radioactive decay than through chemical reactions.

As radioactive decay is a process occurring over many hundreds or even thousands of years, the useful lifetime of an RTG is much longer than that of a battery or fuel cell. This fact has allowed scientists to extend missions powered by RTGs beyond their planned duration. The *Voyager* spacecraft, for example, were only intended as missions to Jupiter and Saturn. However, because the spacecraft performed flawlessly, and there was available power, the missions were extended. *Voyager 1* was diverted out of the plane of the solar system to study an unexplored region of space. *Voyager 2* continued on to study Uranus and Neptune, finally moving into interstellar space. Twenty-five years after the *Voyager* spacecraft were launched in 1977, they are still sending back information, thanks to the power provided by the heat of radioactive decay. It is the decay of plutonium-238 that provides this heat, and, in the minds of many, therein lies the problem.

Plutonium, like most elements, comes in a number of different forms called isotopes, where an atom of a particular isotope is distinguished by its mass. Different isotopes of the same element have different atomic masses. There are fifteen isotopes of plutonium, but only two have real technological interest, plutonium-238 and plutonium-239. The numbers 238 and 239 indicate that the isotope is 238 or 239 times as heavy as normal hydrogen. Though having different masses, different isotopes of the same element are characterized by almost identical chemical properties. They may have radically different nuclear properties, however.

Chemical properties describe the manner in which the atoms of elements combine to form compounds. For example, both plutonium-238 and -239 combine with oxygen to form a ceramic with a ratio of oxygen to plutonium atoms of two to one. Nuclear properties, on the other hand, are concerned with the transmutation of one element into others, often accompanied by a liberation of energy.

Plutonium-238 and -239 decay by a process called fission, in which lighter atoms are formed from a single heavy atom. In the case of plutonium, the fission products are uranium and helium (the helium nucleus is also called an alpha particle, with a mass about four times that of a normal hydrogen atom). In this decay process, which is occurring all the time, energy is liberated in the form of heat. However, because at any given time only a very small fraction of the plutonium atoms are undergoing radioactive decay, the amount of heat generated need not be overwhelming. A sphere of plutonium-238 or -239 weighing a few ounces would be warm, perhaps quite hot to the touch, but it would not get hot enough to melt or to ignite a fire. In a sphere this size, heat is being carried off to the surroundings as fast as it is generated. However, a sphere made from six or seven ounces of plutonium-238 is a different animal altogether. Here, more heat is generated than can be radiated through the surface, causing the sphere to heat up until it melts—at about 1,000° F.—when it can adopt a shape with a greater surface to volume ratio and cool down.

With plutonium-238 all you get is heat, helium, and uranium. Plutonium-239, however, has an interesting additional property: It is a fissile material. When hit by a neutron it can be made to fission, and split into two lighter atoms, in the process giving off two neutrons and energy—mostly in the form of X rays. If each of these neutrons were to hit different plutonium-239 nuclei, each of these would fission, giving off twice as many X rays and four neutrons. If the sample were of the proper shape and size (a sphere weighing several pounds), so that the probability of a neutron hitting a plutonium-239 nucleus were high, eight atoms would fission, followed by sixteen, thirty-two, sixty-four, etc. In short order, nearly every atom in the sample could be induced to fission and liberate a tremendous amount of energy, most in one short burst of X rays.

Most of the X rays liberated by fission are quickly absorbed by electrons, for a short time tearing them loose from the nuclei of nearby atoms. As these free negatively charged electrons recombine with positively charged nuclei, a cascade of radiation is released, sometimes so powerful that it penetrates deep into tissue, liberating its energy in the destruction of biological molecules. Less energetic radiation is the blinding white flash accompanying a nuclear explosion. Still less energetic radiation heats the surrounding air, causing it to expand violently and creating a shock wave moving outward at near the speed of sound. So forceful is this shock wave that it rips trees from the ground and buildings from their foundations. Herein lies the destructive power of fissile materials, making possible bombs, like those dropped on Hiroshima and Nagasaki at the end of World War II.

Even though plutonium-238 is not fissile and cannot spark or take part in a chain reaction, the constant supply of heat from radioactive decay can be converted to electrical current. With plutonium-238 supplying the heat, an energy supply without moving parts is easily constructed, providing a simple, almost fail-safe power system, just the thing needed for robotic spacecraft costing billions of dollars and operating outside the service area of Electricians' Local 454.

There is just one trouble. Too many people associate "atomic bombs" with all isotopes of plutonium. If I were a resident of Cocoa Beach or Cape Canaveral, Florida, I would be very concerned if NASA were launching a satellite containing seventy-two pounds of plutonium-239. But I would be comparatively unconcerned to know that just a few miles away plutonium-238 was being hurled into the sky by millions of pounds of explosive rocket fuel. Many people do not recognize the difference.

One anti-*Cassini* website, "*Cassini*: A Poisonous Package for Saturn," suggests that a nuclear explosion results when too many radioactive atoms are brought together. The fact that this is only true for fissile materials is totally ignored. Given the misinformation, it is not unreasonable to suspect that some of those protesting the *Cassini* launch were doing so because they were genuinely concerned that there was some probability of a nuclear explosion. Though these fears were unfounded, there were other risks posed by the launch that did, indeed, merit concern.

Plutonium has been called the most toxic substance known to man. This reputation is undeserved. While plutonium should not be handled carelessly, it is far less dangerous than many naturally occurring compounds. If forced to choose between ingesting plutonium or botulinum toxin, I would go with the plutonium. More than likely, I would never experience a day of ill health as a consequence of my snack. The same cannot be said of the alternative decision.

Botulinum toxin is a neural toxin produced by *Clostridium botulinum,* a bacterium found in soil and on fruits and vegetables, as well as on meats and fish. For the most part there is little to worry about from this bacterium, for it does not produce toxin in air. But place it in a moist, oxygen-free environment and *Clostridium botulinum* is one happy camper, producing an extraordinarily powerful toxin as it grows and reproduces, a teaspoon of which, if properly administered, is sufficient to kill 100,000 people.

"So what?" you might ask. "We live on a planet rich with oxygen and have no need to fear *Clostridium botulinum.*" Well, that's true as long as we don't inadvertently provide the bacterium with the conditions it needs for growth, which every so often is done while canning fruits and vegetables. If the goods to be canned are not heated to high enough temperatures to kill the spores of the *Clostridium botulinum* bacteria, once sealed in the oxygen-free mason jars they will grow, lending a deadly aftertaste to grandmother's wax beans. Despite the potential for mass poisonings, I have yet to hear of protestors seeking an injunction to stop the neighbor lady's fall canning.

Why is it that botulinum toxin does not evoke the same visceral response as plutonium? I have a theory to explain this observation, based on the way people categorize knowledge.

When confronted with new information, we place it into one of three categories: things we understand, things we don't understand but want to, and things we don't understand and probably never will. Information in category two is dynamic, and when we are young it is this category that is the most active. Everything is new and we have no context allowing for a further categorization. So the information is filed at the front of our minds where we can pull it out easily. This we do over and over again. At some point in our mullings we may come to understand our observations, in which case they are placed into category one. If we are unable to make sense of the information, it is either returned to category two to be considered again or we give up and consider it unfathomable.

Unfortunately, too often as we grow old, information spends less and less time in category two and goes directly into category three. This is why children may ask the question "Why is the sky blue?" but adults almost never ask it. Children are striving to move information from category two to category one. By and large, adults have given up, filing this observation in category three.

Once information is categorized as comprehensible or incomprehensible, it is further grouped and characterized.

While this process proceeds logically for information in the understood category, associations determine how information in the incomprehensible category is characterized. For example, for most people botulism, and, for that matter, all manner of food poisoning, falls in the understood category. Food poisoning is caused by bacteria, which can be killed by properly preparing food—most notably by heating the food sufficiently. That in the United States food poisoning results in 76 million illnesses, 350,000 hospitalizations, and 5,000 deaths a year seems less significant than is the knowledge of what causes food poisoning and how to control it. In contrast, plutonium falls into the not-understood category where it is characterized by association.

In an informal word-association game, 85 percent of the people I questioned responded to plutonium with the word *nuclear*. To *nuclear*, the top three responses were *bomb*, *war*, and *winter*. That in the United States the number of deaths resulting from plutonium exposure or nuclear accidents can be counted with fewer fingers than are on one hand is not as apparently important as those images with which plutonium is associated.

As a further example, many people have benefited by the latest in medical imaging technology. One such technology is MRI, or magnetic resonance imaging, which, among other marvelous things, allows skilled radiologists to find tumors that were previously unobservable. The technology, however, is not new; it has been around for nearly half a century. It was, and still is, used to determine how atoms are bound together to form molecules, and is referred to by those who use it for this purpose as NMR, for nuclear magnetic resonance. Developments in electronics and information processing allowed NMR to become MRI. But why change the name? Why not just call it nuclear magnetic resonance imaging? The answer is that not many people know how MRI works. So MRI is placed in the not-understood category of knowledge, where decisions are made based on associations. Many patients, told to lie down in a nuclear magnetic resonance imaging machine, would associate nuclear with "bomb," and that would be that—a lot of NMRI machines would be sitting around unused. But call it a "magnetic resonance imaging" machine and the patient associates it with those little horseshoe-shaped things he played with as a kid, and in he goes.

Returning to the main topic: Before making an informed decision whether *Cassini* should or should not have flown, it is important to understand the true risks posed by the heat source.

It is the alpha particle (helium nucleus) ejected by plutonium that is responsible for its toxic effects. Because energy is conserved, as high-energy plutonium decays to lower-energy uranium and helium additional energy must be liberated. This energy is initially found in the motion of the newly formed atoms. Of these, it is the alpha particle with the larger part of the decay energy and therefore the greatest velocity. The alpha particle transfers some of this energy of motion to nearby atoms as it smashes into them. They, in turn, smash into other atoms and transfer a little energy. The process continues until the excess energy derived through fission is spread among many, many atoms, slightly increasing their velocities. We perceive this as an increase in the temperature of those atoms.

It is the first few impacts by the alpha particle that are of greatest concern from a health perspective. For if the alpha particle hits a cell, it can, and often does, damage it. If the alpha particle is moving fast enough, it will not be stopped by this first impact and will move on to damage more cells. Fortunately, the alpha particle emitted through the radioactive decay of plutonium penetrates only a few cells.

Of the three routes to plutonium exposure—contact, ingestion, and inhalation—only one poses a serious health risk. Because a piece of paper or the outer layer of skin, where the cells are already dead, will stop an alpha particle, simple contact with plutonium or its dust should not cause undue concern. Ingestion of small quantities of plutonium is also unlikely to cause long-term health risks, since only one atom of every 10,000 consumed remains in the body for any length of time. However, inhaling very fine particles of plutonium is another matter. These particles may lodge in the lungs and remain there for protracted periods. Over time, the emitted alpha particles damage surrounding cells, causing some to become cancerous. Inhalation of plutonium dust is the most serious and credible threat to health associated with the use of plutonium-238 in radioisotope thermoelectric generators. The construction of *Cassini*'s RTGs was intended to minimize this risk.

Two principles were brought to bear in the design of RTGs. First, the plutonium had to be contained in a package that resisted heat and shock. Second, against the possibility that this package would rupture, the plutonium should be in a form that would not easily break into fine particles, which might be inhaled and lodge in the lungs. The second of these principles was the most straightforward to apply.

Plutonium metal combines readily with oxygen. Though I have never seen this reaction occur, it is probably every bit as spectacular as when other metals burn. Magnesium burns with an intense white flame, and is a common addition to fireworks. Titanium also burns readily, and must be welded in an oxygen-free atmosphere. Plutonium and uranium burn so readily that precautions must be taken to avoid sources of ignition. Accordingly, these metals are worked with copper tools, because copper does not produce fire-starting sparks when struck, as do steel and iron. The product of plutonium combustion is, of course, plutonium oxide ash. If conditions were right, much of this ash would be of just the right size to lodge deep within the lungs of any person who was unfortunate enough to inhale at the wrong time. To eliminate this possibility, the plutonium in the RTGs had to be in an unburnable form.

Unlike potatoes that can be twice-baked, plutonium cannot be twice-burned. For the most part, once an oxide is formed, it cannot be further oxidized, making plutonium oxide an unburnable form of plutonium. The plutonium in *Cassini*'s RTGs is in oxide form. This ceramic form of plutonium still undergoes radioactive decay, but now will not burn and, except in the most severe of impacts, will break into pieces too large to be inhaled.

The seventy or so pounds of plutonium used to power *Cassini* are assembled from what are called general-purpose heat sources, or GPHSs. These were designed to further reduce the risks posed by plutonium-238. Each GPHS contains four marshmallow-shaped plutonium oxide fuel pellets about 1.3 inches long by one inch in diameter, and weighing about 1.4 ounces. Each fuel pellet is encapsulated in an iridium container.

Unlike steel or even gold and platinum, iridium is one of those metals you don't see every day. You cannot walk into a jewelry store, even Tiffany's, and find an iridium engagement ring. Not that it wouldn't be beautiful. It looks much like platinum but with a slightly yellow hue. Though very rare, it can be had at a price comparable to other precious metals such as gold, palladium, and platinum. The reason iridium is seldom or never used in jewelry is that it is difficult to work—very hard, very dense, and very tough. Tough as it is, it can be made tougher by adding trace amounts of thorium. Though no one knows exactly how thorium improves the toughness of iridium, the resultant alloy is ideal as the skin of an impact-resistant package for plutonium marshmallows. Iridium's other attractive properties include a high melting point—nearly 2,500° C., or about 4,500° F.—and extreme corrosion resistance. Even the most corrosive acids have no effect on iridium.

As an additional safety measure, a high-strength graphite cylinder, called a graphite impact shell or GIS, further contains each pair of iridium clad plutonium oxide marshmallows. This shell is intended to minimize the damage to the fuel pellets from free fall or impact due to explosion. Next, pairs of impact shells are packaged in a roughly four-by-three-by-two-inch container to form the GPHS. This container, known as an aeroshell, was designed to withstand the heat of reentering earth's atmosphere. Finally, eighteen GPHSs are assembled to form one RTG. *Cassini*'s power is supplied by three RTGs generating approximately 850 watts.

The nested containment design of the RTGs has been extensively tested to ensure that plutonium-238 will not be released in an accident. Simulating possible mishaps, the modules have been blown up, heated in furnaces, hit by projectiles, and exposed to seawater. In each case the radioactive fuel was contained, or at worst there were inconsequential releases. In addition to tests and simulations, there are real-world examples attesting to the rugged construction of RTGs.

In 1968, a weather satellite was destroyed by a launch vehicle malfunction. The spacecraft's two RTGs were recovered intact off the coast of California. The plutonium from these heat sources was reused in other RTGs.

As dramatized in the movie starring Tom Hanks, *Apollo 13* was to land on the moon and deploy various experimental packages to be powered by an RTG. Unfortunately, on the way to the moon a fuel cell providing power for the mission exploded, crippling the command module. Luckily the lunar and command modules docked before the explosion. It was the lunar module that provided the power necessary to return the crew safely to Earth. Instead of remaining on the moon, however, the RTG attached to the lunar module reentered the earth's atmosphere without release of plutonium. Today the RTG remains intact, peacefully heating water at the bottom of the Tonga Trench, deep in the Pacific Ocean.

Iridium-wrapped, graphite-clad, aeroshell-encased plutonium oxide—every structural component of an RTG was designed to prevent plutonium from turning into dust during some cataclysmic event. In a very real sense, RTGs reflect the ultimate in fracture-resistant technology. It might seem that they are leakproof, but they are not. If sufficient heat or force is applied, every structural material will fail. There is no way around it. And though the RTG of *Apollo 13* survived the heat generated from the 20,000-mph reentry of the lunar module, no one believed that *Cassini*'s RTGs could survive a 40,000-mph reentry, which is how fast the probe would be going as it made its Earth fly-by. A slight deviation from the planned trajectory, and *Cassini,* along with its three RTGs, would disintegrate in the Earth's atmosphere. What would be the consequences? Well, here is where scientists have a tendency to get themselves into trouble.

Starting in the 1950s, as the transistor began to replace vacuum tubes, a new scientific tool evolved. That tool was the computer, which enabled scientists to reduce vast amounts of data to a comprehensible form. As computers became faster and more powerful, the "computational" disciplines emerged. Universities began offering classes with names like "computational physics" and "computational chemistry," and eventually one could major in these fields—earning an advanced degree in computational "whatever." Because there were two parts to these particular specialities—the computational part and the "whatever" part—both those interested in computation and those interested in "whatever" were attracted to these fields of study.

For those interested primarily in computation, success was measured in terms of how fast and how accurately larger and larger quantities of data could be reduced to a manageable form. Less emphasis was placed on what these reduced data had to tell us about the field in question. Over time the practitioners of computational "whatever" diverged into two groups: Those more concerned with computation became modelers, while those more concerned with "whatever" became simulators.

The differences between modelers and simulators are not generally recognized within the scientific community. Though there might not be a bright line separating the two, I believe the distinction is justified. A modeler uses computers to answer the question "what if." For example, what if atoms behaved as tiny ball bearings—feeling the presence of other atoms only when they touched? How would materials made from these atoms behave? To answer this question, a modeler would use computers to assemble billions of these atoms into a virtual material and explore its properties. On the other hand, a simulator would be uninterested in this problem, because real atoms do not behave like ball bearings. Real atoms interact with each other in very complex ways. The simulator would prefer to treat these interactions as accurately as possible, even if this made it impractical to investigate more than a single molecule made from a few "real" atoms.

In recent years, technological advance has resulted in part from simultaneous advances in both modeling and simulation. One such advance derives from the ability to simulate the mechanical properties of structures such as bridges, buildings, cars, and planes. This is accomplished by assuming these structures are composed of thousands of little "elements" as small as a few millimeters on a side. Thanks to simulators using the principles of fracture mechanics, the mechanical properties of each element can be accurately predicted with the aid of computers. Modelers have provided the necessary formalism to couple many thousands of these elements together. Combining these capabilities, it is now possible to simulate the mechanical behavior of structures that have yet to be built. Now the engineer can test various structural designs without needing to construct prototypes, an unheard-of practice just twenty-five years ago. Eliminating this need has acclerated the pace at which new products are introduced. Boeing's 777, for example, was designed and produced in only a fraction of the time previously required for the introduction of a new airplane, simply because computer-aided design provided a feasible alternative to conventional build-and-test. The same can be said of products other than planes. Now a new automobile can be designed, built, and marketed in less than a year.

Though the true benefit to technology from the computational sciences has come from the combined efforts of simulators and modelers, I am willing to bet that most people lump the two together, assuming that they do essentially the same thing. I do not believe it is generally recognized that modelers are extending the boundaries of computation, and that their models need not correspond to anything real. For example, a multitude of models purport to predict the effects of pumping carbon dioxide into the atmosphere. The first step in constructing such models is to guess at the underlying governing principles. Carbon dioxide traps heat, so Modeler A designs an efficient algorithm to determine how this heat will move through the atmosphere and change global temperatures. The challenge to the modeler is resolution. A very crude model might predict the average temperature change over the eastern United States, while a more sophisticated model will predict the temperature change in Boston. The finer resolution requires more calculations and hence a bigger, faster computer. This computer crunches away on countless calculations, and—lo and behold—Boston warms up. Previous models only predicted that the eastern United States would warm up. In accordance with time-honored protocol for scholarly research, Modeler A publishes the results of his model in a journal catering to this type of research.

As is typical, one of Modeler A's students seeks to extend her mentor's work. "Wait!" she says. "Won't the higher temperatures mean that there is more water vapor in the air, making white, billowy clouds that will reflect sunlight back into space?" She makes the model more computationally complex by including reflectivity equations. A new, even faster computer crunches away and—lo and behold—Boston cools down. The results of this model are once again published, and everyone is impressed that the model is getting more complex. Modeler B, at another university, reads this paper and says, "Wait, the effect of carbon dioxide dissolved in seawater has not been considered." Modeler B adds more complexity, crunch-crunch goes the newest, fastest computer, and—*voilà*—Boston is now unchanged, neither heating nor cooling.

The point is, models are just that: models. Most are constructed to take full advantage of the increasing power of computers, not to reproduce reality. This is a much more difficult problem that cannot be solved until simulators understand underlying mechanisms, and until modelers have developed the tools to treat the complexity arising therefrom.

So what has all of this to do with what would happen if *Cassini* were to reenter the atmosphere during its Earth fly-by? Well, no one knows, so the scientists in charge of writing the environmental-impact statement turned to modeling. In one scenario, about 30 percent of *Cassini*'s plutonium would be lost to the atmosphere and spread over 2,000 square kilometers, causing 2,300 cancer deaths over fifty years. In a supplemental environmental-impact statement, the area of coverage was reduced to about eight square kilometers, causing 120 deaths over fifty years. The obvious uncertainty gave the anti-*Cassini* forces a good deal of ammunition. Which was it, 120 deaths or 2,300 deaths, or maybe some much larger number? Shouldn't the launch be halted until we know for sure?

Not in the foreseeable future will an experiment or model tell us how many people would die from cancer if *Cassini* were to burn up in the atmosphere. The myth that complex phenomena can be simulated does a disservice to both science and the public. What science can do is provide a framework to assess relative risks.

When I awoke on the morning of August 15, 1997, I was confronted with many potential hazards. If I had considered these and ranked them from most probable to least probable, the list would look something like this:

1. Suffering an injury while riding my bicycle during lunch hour.

2. Being in an automobile accident while commuting either to or from work.
3. Having an accident at home—slipping in the shower or falling down the steps.
4. Daily exposure to toxins from the polluted air around Denver.
5. Exposure to second-hand smoke.
6. Exposure to radon.
7. Eating fatty foods.
8. Becoming a victim of crime.
9. Having a work-related accident.
10. Being struck by lightning.
11. Getting caught in a flash flood.
12. Being injured by a tornado or high winds.
13. Suffering an injury from an attack by a mountain lion or bear.
14. Inhaling plutonium from the reentry of *Cassini* during Earth fly-by.

I am not opposed to minimizing risk, but if I am to approach the problem of minimizing risk as a scientist, those things that pose the greatest risk to life and health should receive the greatest portion of our attention. Which means I, for one, should probably stop riding my bicycle—and that, dear reader, I will not do.

The RTGs aboard *Cassini* and those that will be used to power other space probes are well designed to resist breaking. They are tough. In my opinion, they are tough enough to go.

WHY ASK WHY?

CHAPTER

9

Knowing when something will break has served us well. We can make incredibly tough containers, amazingly strong sheets of glass, ceramics that shred shrapnel, and polymers that absorb the energy of bullets. In less than a century we have moved from a society where breakage was considered the norm to one where failure to design things that don't break may result in billions of dollars in civil liabilities. It is expected that the knowledge of when things break will be used to design safer cars, planes, and ships. Our tolerance for things breaking is vanishing simply because the demands we place on those things are increasing.

To make our world safer while simultaneously improving the "things" of the world, understanding when something breaks no longer suffices. It will be necessary also to understand why things break.

Among the "things" most in need of improvement are those that generate or utilize energy. And among those, heat engines are at the top of my list. It's not a common term, but heat engines are very common in the modern world. Indeed, they constitute a broad category including everything from internal combustion engines found in cars and trucks to the turbines powering jet aircraft. Other turbines, which convert the chemical energy of oil or natural gas into electricity, also fit into the category.

As their name implies, heat engines derive power from heat. Usually this heat is converted to work through the expansion of a gas. For example, in a common internal combustion engine, gasoline vapors are drawn into a cylinder where they explode, generating heat that causes the combustion gases to expand. The expanding gases move the piston. It is this motion that is harnessed to do useful work.

How much work can be extracted from heat is a measure of an engine's efficiency. In a 100-percent efficient engine, all heat would be converted to work. The common belief is that it is impossible to achieve 100-percent efficiency because of friction. The story on the street is that in an ideal world, where friction has been defeated, where perfect lubricants are available at the neighborhood hardware store, where machines are designed and manufactured with total accuracy, with every bearing of exactly the same size and perfectly spherical, 100-percent efficient heat engines are attainable. In this world, cars deliver hundreds of miles per gallon, and oil is as cheap as dirt. Unfortunately, the reality is that even in a friction-free world, 100-percent efficient heat engines are impossible.

It was the French engineer Sadi Carnot who made this remarkable discovery in the early part of the nineteenth century. In the process he also uncovered the Second Law of Thermodynamics and the concept of entropy. Carnot was studying the efficiency of the newly developed steam engine—one of the earliest incarnations of the heat engine. Carnot knew that energy is conserved—remember, this is the First Law of Thermodynamics. So, as a gas expands and loses energy to work, that energy must come from somewhere, which in this case is the heat energy of the gas. In order to get work out, the gas must cool as it expands. The gas starts out with a temperature we will call T_H, and as it expands, it cools to a temperature we will call T_C. As an example, for an internal combustion engine, T_H is the temperature of the cylinder gas immediately after being ignited by the spark plug, and T_C is the temperature of the exhaust gas at the exhaust valve. Carnot showed that the maximum attainable efficiency of a heat engine is $1 - T_C/T_H$.

Let's look at this equation in a little more detail. For our example, if the ignition temperature is twice the exhaust temperature, $T_C/T_H = ½$. So the maximum efficiency of the engine is 50 percent. This is the efficiency of the engine in our ideal friction-free world. In the real world, plagued by problems of friction and other imperfections, its efficiency will be less. In order to get the engine efficiency up, the ignition temperature, T_H, must increase and/or the exhaust temperature, T_C, decrease. Of these two options, it is easier to increase T_H. But as T_H increases, so will the temperature inside the heat engine. The conclusion is inescapably simple: To squeeze more work from a fixed amount of fuel, heat engines must operate at the highest possible temperatures. So what prevents us from constructing heat engines that operate at temperatures ten times present values? Well, nothing—except that at this point we simply don't have materials with the right combination of properties.

Consider your typical family car. There is very little preventing us from using a fuel that will burn much hotter than the 87-octane stuff dispensed at the local gas station. One possibility is nitro-methane, the fuel used by drag racers to propel their cars from zero to 300 miles per hour in six seconds. Part of what makes this possible is the tremendous temperature at which nitro-methane burns, increasing the engine efficiency but also increasing operating temperatures. We could, with the help of a good mechanic, design a system to inject a little nitro-methane into the engine of the family car, and the sedate sedan would be filled with awesome potential. For a brief period Ferraris and Corvettes would be left in the dust as the innocuous-looking auto peeled away at stoplights. I say brief, because after a few jackrabbit starts, the metal exhaust valves of our nitro-methane-burning automobile will have melted or oxidized away.

Okay, undaunted we continue our project and develop new exhaust valves specially made from an alloy that resists oxidation; so off to the races. Once again Ferraris and Corvettes are hapless victims of the more efficient engine. Until the fifth or sixth stoplight, when from under the hood comes the horrific grinding of broken machinery.

When the mechanic finally opens the engine up, every one of the new valves is found shattered. Some of the pieces have fallen into the cylinders, scarring the walls. Lifters are bent, rods deformed; the engine is a total loss.

The new valves broke because the stuff from which they were made was brittle. The original valves failed because the stuff from which they were made oxidized. The solution seems obvious. We need a tough material with a high melting point. Scanning the literature looking for such a material, we discover the following. Of the tough materials, all will melt or oxidize when the temperature reaches the operating temperature of our nitro-methane-burning car. Of the high-melting-point materials, all will break when subjected to the forces and thermal stresses of our more efficient engine for more than a few seconds.

If we are still determined to burn nitro-methane in our car and expect it to run reliably without a complete engine rebuild after just a few minutes, or seconds, of operation, there are two options. First, the valves could be redesigned. The metallic valves could be coated with a ceramic layer, insulating them from the heat. Or we could try to circumvent the oxidation problem. Unburned fuel, circulated around and through the exhaust valves, would act as a refrigerant, cooling the valves. The second option is to find a new material that is both tough and able to withstand the environment of exploding nitro-methane.

Both options have their drawbacks. The first relies on design to compensate for less-than-optimal material properties. As this option is exploited, the "things" of the world are becoming more complex. And as a thing becomes more complex, there are more ways it can fail. Reliability and complexity are incompatible—though this simple truth is often unnoticed as improved manufacturing and fabrication techniques offset the loss of reliability. The second option—to find a new material with better properties—is plagued with conceivably the most serious drawback: It might just be a waste of time. What if there is no such material? The history of science is rife with fruitless searches. How much time was expended in the alchemists' doomed efforts to convert lead to gold? Even the illustrious Sir Isaac Newton was caught up in this search. Would humanity have profited had Newton used his time more constructively? We will never know.

Had the alchemists known why certain combinations of substances changed form—that brittle green malachite, when heated, transformed to ductile red copper through a rearrangement of atoms—they would have known that these transformations did not imply that one kind of atom could be converted to another. Knowing the "why" of such a question supplies information about what is possible . . . and what might just be *im*possible, too.

Is it possible to make a tough material that is stable at high temperatures? No technology is more reliant on an answer to this question than aeronautics, where the search for a new generation of materials from which to build jet turbines is proceeding at a fevered pace. Jet turbines are the heat-engine power plants that propel aircraft to supersonic—even hypersonic—speeds. As heat engines, the hotter the turbine runs, the better.

The first jet turbines were constructed from stainless steel. But steel, even stainless, reacts with oxygen at fairly low temperatures. So, during the 1950s a less reactive material, nickel, became the principal metal of high-temperature components in jet turbines, which at that time operated at temperatures around 1,200° F. The arms race of the Cold War supplied constant upward pressure on those operating temperatures. And with each year that passed, operating temperatures climbed about 25° F. This constant increase was brought about through incremental improvements to high-temperature nickel-based alloy, its processing, and turbine design.

Today's jet turbines look nothing like those of the mid-twentieth century. The turbine's whirling compressor blades are formed from single crystals of nickel alloyed with chromium, aluminum, titanium, cobalt, boron, carbon, molybdenum, tantalum, hafnium, iron, and tungsten. Every ingredient in this metallic stew confers specific properties that, taken together, transform mild-mannered nickel into a nickel-based "superalloy"—that is what they call it. Each superalloy blade is covered with a "self healing" ceramic thermal barrier coating, a skin that insulates and protects the underlying metal from oxidation. If scratched, the blade literally grows new skin. These highly engineered alloys operate continuously at temperatures near 2000° F. and for short periods substantially higher. So close to nickel's melting point (about 2,600° F.) there is little probability of brittle fracture. At these temperatures the problem is creep, a phenomenon in which a material deforms under load, much as Silly Putty elongates as you spin it about your head. These superalloys, however, have to be dimensionally stable and operate at extreme temperatures for thousands of hours without changing shape.

While incremental increases in operating temperature have sustained the aircraft industry, there has been an ongoing search for a new material that will enable a quantum leap in jet turbine efficiency. This material, if it exists, will resist creep and oxidation at temperatures above 2,500° F., where nickel-based superalloys have the consistency of butter. Maybe an alloy based on the element niobium will be the ticket. Niobium melts at almost 4,500° F., but it never gets this hot in air, as it has this rather nasty habit of first bursting into flames. Perhaps there is some way to coat niobium to protect it from oxygen. If so, this coating had better self-heal and do it fast. If even a tiny scratch were to expose niobium to air at 2,500° F., it would soak up oxygen faster than Bounty—"the quicker picker-upper"—soaks up water. But maybe just a little bit of aluminum or chromium, or a combination of the two, would provide the necessary self-healing.

"Maybe," "perhaps," "if"—words used when we don't know *why*. We don't know why niobium melts at 4,500° F. while nickel melts at 2,600° F. We don't know why nickel resists oxidation while niobium does not. Melting, oxidation, and breaking are phenomena that are controlled by the chemical bonds holding stuff together, and we simply don't fully understand how those bonds work. So, when it comes to finding new materials with properties derived from bonds, we do it the old-fashioned way—we use trial and error.

As I started my graduate studies in the fall of 1979, the agencies of the U.S. government, supporting much of this country's basic research, recognized that finding new materials by trial and error was too slow and much too costly. Then there was that nightmare: Some of the materials researchers were looking for might not exist. Spending years and years looking for materials with an impossible combination of properties would strike some (read voters and taxpayers) as wasteful. And no government agency wants to be seen as wasteful by those elected officials who control its budget. And elected officials don't want to be seen as supporting wasteful projects. (Unless, of course, they are supporting wasteful spending in their own district, but that's another story.) So an alternative approach to finding new materials was necessary. Developing this approach became a high priority throughout the funding community.

In general, funding agencies support basic research as follows. First, a "call for proposals" (CFP) is issued. This call asks prospective investigators to describe a research program that is likely to provide the capabilities or objectives outlined in the CFP. This description—the proposal—summarizing the research approach and anticipated results, along with projected costs, is returned to the funding agency. There it is evaluated for merit and, in principle, the most meritorious proposals are awarded. The recipients of these awards are termed principal investigators (PIs). They spend the next three years or so pursuing the proposed research, all the while communicating with the granting agency as to progress. In the United States, most of the research conducted at universities is supported via this mechanism.

The stage was set. In the late 1970s and early 1980s, the policy makers, in order to establish broad directions for the funding agencies, were looking to jump-start a new approach to materials discovery. They didn't know what this new approach would involve; they just wanted it to be less reliant on trial and error, more purposeful. The term "materials design" seemed to capture the essence of the purposeful hunt for stuff with novel properties. The first CFPs would call for new capabilities to support an emerging field of materials design. What exactly "materials design" would consist of was not clearly stated, basically because the policy makers had no more than a blurry vision of the field they wished to create. However, there was no doubt that the new field would involve computers, evidently to determine the properties of materials that had yet to be made. In the extreme, it seemed that funding agencies were hoping materials design would ultimately allow an engineer to specify the properties of materials required for particular applications. These would form the inputs to some elaborate materials design software: Chug-chug, and out comes not only the composition, but also a processing regime to make the needed material. If no material could be made with the desired combination of properties, then the computer program would identify a substance with the properties closest to those requested. These grandiose objectives were, at best, very long-term. In the short run, the funding agencies had to develop more modest and achievable goals. What should they be?

It is 1979, and Greg Olson at MIT is pondering the problem of embrittlement. (Remember the *Titanic*? A little sulfur, and normally tough ductile iron shattered on impact with that iceberg.) Greg is asking himself why sulfur embrittles steel while boron does not. He reasons that it has something to do with chemical bonds and suspects that Slater and Johnson's computer codes provide the means to study these. Aware that the funding agencies are looking for short-term goals for their materials design programs, Greg did what effective scientists do. He set about helping the agencies "see" their way.

Both Greg Olson and Keith Johnson worked closely with researchers at the General Electric research facility in Schenectady, New York. General Electric is the largest producer of gas turbines (yup, heat engines) in the world. GE makes turbines that generate electricity and power jets, both military and civilian. If you fly commercially, GE engines probably get you where you are going. And that electricity used to run your computer and light your house—almost certainly a large part of it was generated by GE gas turbines.

GE became the world's largest producer of turbines by making an efficient product. This efficiency comes from carefully balancing design with reliability. And that balance can be achieved only when the materials from which the turbine is constructed possess highly consistent properties. It just would not do to build the blades for a jet turbine from an alloy that had a 5-percent variation in strength over time. To compensate for this variation, the designer would need to make heavier blades and sacrifice efficiency. In turbines, as in most other products, an optimal design requires materials with consistent and unchangeable properties.

Consistency of properties is one of the main problems confronting designers. This is particularly true for turbines operating in constantly changing environments, which can produce corresponding changes to materials' properties. Not only is a turbine subject to dramatic thermal and stress cycling, but also to hostile and constantly changing chemical environments. Impurities in the air or fuel may expose jet turbine blades to deleterious elements, reducing toughness. The consequences could be catastrophic.

Even if researchers and designers at GE were to discover a new alloy with superior toughness at high temperatures, more-efficient turbines would not follow immediately. First would begin a protracted period of empirical investigation to answer very pragmatic questions. Of all possible impurities, which are likely to embrittle? Is there a combination of thermal and stress cycles that might lead to materials failure? The only way to answer such questions is to test every possible scenario—a lengthy process. This is why the time between the discovery of a new material and its commercial use can be as long as twenty-five years. Obviously, speeding up this process is a high priority, particularly at innovating companies like GE.

Commercial labs like GE's, in addition to responding to CFPs, also fund research with company profits. Much of this "internally supported" research is intended to meet the needs of profit-making divisions. GE's aircraft engine division, for example, may have a problem with a manufacturing process and turn to its research labs for assistance. However, a small portion of the research budget is also used for high-potential but also high-risk projects. Most often these are entrusted to senior investigators with a proven research history. Olson and Johnson's collaborators at GE were such investigators. Clyde Bryant is a prominent metallurgist who, in 1979, was working at GE along with Dick Messmer, a quantum chemist and friend of Keith Johnson.

All four scientists, Bryant, Messmer, Olson, and Johnson, recognized a rare opportunity to do groundbreaking research. Bryant and Messmer would seek internal funding from GE to support the first quantum mechanical investigation exploring why certain elements embrittled while others did not. Johnson and Olson would seek funding from the National Science Foundation (NSF) to support a young graduate student at MIT involved in similar research . . . and just in case you didn't know, that graduate student was me.

On the one hand, GE supported the proposal of Bryant and Messmer, while, on the other, NSF declined the proposal of Olson and Johnson. With 20/20 hindsight, NSF's decision was a mistake, which would take twenty years to rectify.

Though I was not funded to explore fracture, both Keith Johnson and Greg Olson encouraged "preliminary" research into the mechanisms of embrittlement. More often than not, preliminary investigations form the grist for future proposals. So, while Bryant and Messmer focused on the mechanism by which sulfur atoms embrittle nickel, I was looking for an alternative system to study, which I stumbled upon in the winter of 1981. At the time I was taking a course in surface chemistry from Ron Latanision, who had an interest in the "hydrogen economy."

For those who think green, hydrogen is often held out as a panacea to the evils associated with burning fossil fuels, most conspicuous among these being greenhouse emissions and global warming. Heat, carbon dioxide, and water vapor are generated when fossil fuels are burned, an extremely useful combination when extracting work from a heat engine. Unfortunately, carbon dioxide is a greenhouse gas, which means that it absorbs radiation to which the other components of the atmosphere are transparent. So energy, which would normally pass through the atmosphere, is captured, so to speak, by carbon dioxide. Some believe that as the concentration of carbon dioxide in the atmosphere increases, the additional absorbed energy will cause global warming. In some models—remember the difference between models and simulations—many of the Earth's more pleasant climes will become sweltering deserts, if carbon dioxide emissions are not checked.

Enter hydrogen. Hydrogen is also a good fuel for a heat engine. When burned, hydrogen produces heat and water vapor, but no carbon dioxide. Hence, if fossil fuels were replaced by hydrogen, we could keep our fuel-guzzling SUVs, which we could then use to escape to the mountains or beaches on hot summer days. Of course, if we continue to use fossil fuels, we can still take those weekend trips, but the mountains and beaches offering relief from the heat will now be in Alaska, Norway, and Siberia. There is also the availability issue, which always comes up when the scientifically uninitiated discuss the advantages of hydrogen over fossil fuels. As I understand it, the argument goes something like this. Hydrogen is the most abundant element on Earth, with every molecule of water containing two atoms of hydrogen. Therefore we can free ourselves from the dependence on foreign oil by converting to hydrogen. What the proponents of this argument fail to understand is that water is oxidized—or "burned"—hydrogen. The energy that makes hydrogen a good fuel has already been released when it was burned. It is indeed possible to recover hydrogen from water, but only by expending energy. Because energy is conserved, at least as much energy is required to free hydrogen from water as will be produced when the hydrogen is burned. From where is this energy to come? Well, at the moment, the only energy source capable of generating hydrogen in the quantities required to power the world's heat engines would be from burning fossil fuels—a conundrum.

Even though the availability argument has serious flaws, replacing fossil fuels with hydrogen is, in my opinion, a good idea. I am less concerned with global warming than by the fact that the United States remains dependent on corrupt and unpredictable foreign governments as long as oil is our primary energy source. Before we can free ourselves from this energy yoke, we must use something other than foreign oil to release the hydrogen from water. How about solar, wind, geothermal, or even nuclear energy? The energy from these sources is easily converted to electricity. In turn, electricity can be used to "unburn" water, producing hydrogen and oxygen. The hydrogen can then be transported to the local gas station via a pipeline, where SUVs can be fueled for those lovely weekend jaunts.

Alternatively, why burn hydrogen in a heat engine at all? The Second Law limits the useful work that can be extracted from heat. Instead, eliminate the first step in the conversion of energy to heat and then heat to work. Use a fuel cell to extract work directly from burning hydrogen. A fuel cell is similar to a battery, where the energy released from a chemical reaction is converted to electrical current and is then used to do work. In principle, 100 percent of the energy available to do work—what we call the free energy—can be harnessed using a fuel cell. In a hydrogen fuel cell, all one gets is water and work—no heat.

Whether burned in heat engines or in fuel cells, many scientists and technologists—Ron Latanision among them—believe hydrogen is simply a more reasonable fuel than carbon-based alternatives. They advocate moving from an economy dependent on fossil fuels to one in which the combustion of hydrogen provides most of our energy needs, a hydrogen economy. This new economy, however, will require the simultaneous development of new technologies to generate, transport, and store hydrogen—but here again, problems arise. Among these, hydrogen embrittles most anything with which it comes in contact—steel, nickel, you name it. So a hydrogen economy also requires the discovery of materials less susceptible to hydrogen embrittlement.

As part of the surface chemistry course, the students were required to do a project. Ron suggested that I use the Slater and Johnson codes to investigate hydrogen embrittlement. Specifically, Ron wanted to know what hydrogen was doing to the chemical bonds of iron to change it from a reliable, ductile metal into an unpredictable, brittle one.

I could not have asked for a better problem, or for a better time of the year to begin my graduate study of fracture. Boston in winter is not pleasant. It is gray and cold and wet. Living in Colorado all my life, I had seldom seen freezing drizzle, and I had never seen an ice storm. Yet in Boston you could not get away from the heavy, icy snow. The ski slopes offered little of the seductive attraction that lured me away from studies in Colorado. On the East Coast, hard-packed means just that—*hard*. For the first time I experienced real pain from even the most controlled of falls. A "groomed" slope had a local meaning differing from what was understood by western skiers as well. In New Hampshire, "groomed" meant the ice had been broken up into marble- to golf-ball-sized chunks. Skiing down these slopes was not unlike driving at 60 mph over a severely rutted backcountry dirt road. I kept a clenched jaw for fear of biting my tongue.

For exercise I took up a local sport—snow shoveling. This was also a new experience. Because it is five degrees farther south and 5,000 feet higher in elevation, Denver receives a higher intensity of winter sun than does Boston. In Denver the sunlight causes snow to turn directly into vapor, a process called sublimation. Essentially, the sun's energy is absorbed by the snow, causing water molecules in the snow to vibrate. The greater the energy absorbed, the larger the amplitude through which the water molecules move. At some critical amplitude, the water molecules break free of the ice and become vapor. Sublimation can occur even when it is well below freezing. On a very cold but sunny day in Denver, the snow of the previous night will just disappear. How fast the snow vanishes depends on how much sunlight is absorbed, which in turn depends on the amount of snow exposed to direct sun. By shoveling it into piles, the amount of snow exposed to direct sun is reduced and does not sublimate as rapidly. So, except in unusual circumstances, Colorado natives simply ignore the snow and it obliges by going away in a day or two.

This is not the case in Boston, where there is insufficient solar energy to sublime snow or ice. Consequently, during protracted cold spells the snow just lies there, getting packed down into ugly, dirty black stuff that coats sidewalks and roadways, making both walking and driving hazardous. That is, unless it is shoveled into piles, then only these large mounds of snow turn into ugly heaps. The roads and walkways remain at least passable, if not attractive.

So, when it snows, the experienced Bostonian rushes out to clear the snow. Pickup trucks sporting snowplows shepherd snowflakes into corrals at the ends of streets and parking lots. The din of snow-blowers fills the air as driveways and walks are turned into cement and asphalt valleys defined by the mountains of snow on either side. But most intriguing are the many box canyons that appear on the street in front of nearly every house. Some of these appear to have been fashioned by true craftsmen who were careful to keep their snow walls plumb and square. At the end of the day, each of these canyons will host a car or two. And as long as the walls of these canyons stand, winter etiquette is in force. Parking in these canyons is reserved for those who constructed them. Occasionally a folding lawn chair or "borrowed" orange traffic cones are used as place-holders during the day, just to remind people of the appropriate code of conduct. By the time I moved to Los Alamos in 1985, I had become quite a snow shoveler, able to shape walkways and parking spots that wouldn't look out of place in a village of igloos.

This discussion of snow and Boston's winter weather may seem extraneous, but it played an important part in motivating my investigation of hydrogen embrittlement. Just like Greg Olson, Ron Latanision felt that a deeper understanding of embrittlement called for a deeper understanding of the relationships between chemical bonding and fracture. However, he took a somewhat different approach toward influencing the funding agencies. He organized a scientific conference on "The Atomistics of Fracture," scheduled for May of 1981. At this conference all the important people working in the field would get together and talk about their work and the directions that future research should take. As with nearly all conferences, the proceedings were to be published. Such published proceedings include papers by invitees and remarks by organizers. These publications not only provide a scholarly record, but also often help to shape the decisions of funding agencies when making policy decisions.

Among those who had been invited to the conference and were to prepare presentations and papers were Keith Johnson, Clyde Bryant, and Dick Messmer, as well as most of the world's experts in the field of fracture. Getting sixty or seventy of the leading researchers in a narrow field together in one place is not an easy matter. It helps to host the conference in a pleasant locale, say a hotel on the beach of a Mediterranean island like Corsica.

As I made progress on my research, Ron and Keith began to discuss the possibility of allowing me to attend the conference. All during that gray, wet winter I thought about little other than hydrogen embrittlement and the reward for solving the problem—Corsica in the spring.

What was the right way to think about embrittlement? I pondered this question while shoveling snow. There was an irresistible temptation to picture chemical bonds as simple connections, like the steel cross-members of a railway bridge. The bridge would fail if the cross-members were somehow weakened. Thus an element embrittled if it weakened the chemical bonds of the base metal. It seemed so obvious, but the Boston snow revealed the flaw in my reasoning.

In newly fallen snow, the flakes are only weakly held together. The most weakly bound flakes are found in the "champagne powder" of the Rocky Mountains. This snow deforms under its own weight. It is not suitable as a building material for a snowman, let alone the walls of a private parking cubicle. Boston's snow is a bit wetter, and the flakes are better held together. It can be molded, shaped, and deformed into marvelous yet utilitarian winter landscapes. But let it sit for even a short period, and the flakes are tightly bound into an icy mass. Forget shoveling; all that can be done now is to break the stuff up.

Progressing from weakly to tightly bound flakes, the snow piles transformed from ductile to brittle. If we imagine the flakes as atoms and what holds them together as chemical bonds, then one might argue that embrittlement results from making bonds between atoms stronger. This too is incomplete, for initially, as the snow became wetter and the flakes more tightly bound, the piles remained deformable.

The key to understanding is not locked up in a single quantity related to the strength of a chemical bond, but in *relative* strengths. In ductile materials, the bonds preventing planes of atoms from sliding across each other break before the cohesive bonds holding the planes together do. In brittle materials it is just the reverse. The question is not how strong or weak the bonds are, but which happens first: bond-breaking allowing sliding, or bond-breaking allowing decohesion. Deformation and fracture can be thought of as competing chemical reactions.

When it comes to two competing reactions, with knowledge of the charge densities (where the electrons are) of reactants and products, chemistry offers a framework to determine which will happen first. In this case, one would compare the charge density of the starting state to that of the charge density as the material begins to fracture or to deform. Whichever is the more similar is the reaction that will be favored. In other words, if the charge density of the material as it begins to fracture looks more like the charge density of the pure material than does the charge density of the deforming material, then it will be brittle. The converse would indicate a ductile material. Under this hypothesis, an embrittling element would shift the charge density of the host metal to make it look more like it does just before it fractures.

The way to Corsica was clear. Calculate the charge density of pure iron. Next, calculate the charge density of iron just as it begins to fracture. Finally, calculate the charge density of iron containing hydrogen and compare all three. If my hypothesis was correct, the charge density of the hydrogen-containing system should look more like fracturing iron than like pure iron. That is what I did, and that is what I saw. Hydrogen shifted the charge density of iron, making it look like fracturing iron.

What happened next was a bit disheartening. I presented a summary of my theory of hydrogen embrittlement to the surface chemistry class in early May. Keith and Ron appeared impressed by the theory and the supporting calculations, and were anxious to have the work presented in Corsica. However, neither could afford the several thousand dollars it would cost to send me to the conference. Ron was already sending another graduate student. And because NSF had declined to support Greg and Keith's proposal, there were no funds available to support my travel to a fracture-related meeting. So, though I did not attend, my work was still represented. Keith gave essentially the same presentation at the conference as I had given to my surface chemistry class. In addition, Keith, Clyde Bryant, Dick Messmer, and I coauthored a paper for the proceedings. The first half reviewed the work on hydrogen embrittlement, and the second presented Bryant and Messmer's work on sulfur embrittlement of nickel. Surprisingly, at least to me, they had fallen into the "trap," suggesting that elements embrittled through the weakening of metal-metal bonds. Nonetheless, I was still honored to have such distinguished scientists as coauthors on my first technical publication.

It was not as if my approach to embrittlement was problem-free. There was, in fact, a weakness, not in the theory per se, but in its application. The problem was in assessing the similarities between charge densities, which is a little like comparing people. Whom do you resemble more, your mother or your father? On occasion there is just no question, but in other cases, it is a judgment call. You may have your mother's eyes and your father's nose, but there is no way to weight these similarities. Chemistry provides an empirically discovered methodology for comparing the charge densities of organic molecules. However, when it comes to comparing metallic crystals, there was no existing formalism. Until I could develop—or find—such a formalism, my theory was of little use. Whether an element embrittled or not would remain a judgment call.

RIGHT ANSWERS, WRONG ANSWERS, AND USELESS ANSWERS

CHAPTER

18

The proceedings of the conference "The Atomistics of Fracture" were published in 1983, more than two years after the participants had departed the sunny beaches of Corsica. This is typical for proceedings. Authors are seldom as punctual as the organizers would wish. Familiar with the delays associated with the publication of conference proceedings, Bryant and Messmer also submitted papers to scientific journals. The journal articles provided more timely publication of their theory of embrittlement, and exposure to a larger audience.

As the articles by Messmer and Bryant began to appear, the funding agencies that had been so cool to the original research proposed by Greg Olson suddenly got it. Seeking the quantum mechanical origins of fracture was decidedly high-tech; it required computers—really big computers at that—and it satisfied an industrial need, otherwise GE would not have supported it. These are all characteristics the funding agencies like to see. Understanding why one element embrittles and another does not became the starting point for the development of the new discipline of materials design.

Almost instantly the CFPs were circulating: Propose research to explain why sulfur segregates to the grain boundaries of steel, thereby embrittling it; propose research to explain why boron atoms increase the toughness of an alloy of nickel and aluminum. The quantum mechanical investigation of fracture, at least one of its aspects, had become a legitimate area of research. More important, there was *money* to be had.

Money thrown at a scientific problem is a little like meat thrown to a pack of hungry dogs—everybody wants a piece. This is because status, promotion, tenure, lab and office space, and so much more are linked to the level of an investigator's research funding. The adage "publish or perish" is incomplete. More accurately, it should be "find funding, then publish to get more funding or perish." Research grants keep department heads, deans, provosts, and university presidents happy because the overhead from these grants helps to run the university. Publications keep funding agencies happy because this is one measure of the effectiveness of the agency and its program monitors. Program monitors work for the funding agency. They review proposals and make the decisions on which investigators will receive awards, and then monitor the research for the term of the award. If monitor Joe gets four technical publications for every $100,000 awarded, while Heather gets six, Joe might find himself looking for a new job when the budget gets tight.

We need to pause here for a commercial message. The Air Force Office of Scientific Research, AFOSR, and the Defense Advanced Research Projects Administration, DARPA, have supported my research for the past several years. I think you will agree that it is fascinating stuff and exactly the kind of thing you want your tax dollars supporting. A letter to your senators and representatives expressing this point would not be out of place. Now, returning to the story:

So, here is this pile of money available to any researcher who has the capability to perform quantum mechanical calculations and is interested in fracture. By virtue of the fact that there is no group of scientists that is perceived to be expert in this arena, no group has an advantage over another. As far as the funding agencies were concerned, solid-state physicists, quantum chemists, metallurgists, and materials scientists were are all equally qualified to perform the research required. As a consequence, funds were awarded by discipline more or less proportionally to the number of proposals that that discipline submitted. In retrospect, it is not at all surprising that the bulk of the proposals submitted came from computational solid-state physics, simply because this was the discipline characterized by both an interest in the solid state, unlike quantum chemistry, and the expertise to perform quantum mechanical calculations, unlike metallurgy and materials science. What the funding agencies failed to anticipate was that a physicist's correct answer to "Why?" may be very different from a metallurgist's or a materials scientist's.

I first appreciated that there could be multiple correct answers to the same question during a high school chemistry class, in which we were using a simple diffraction grating to study the properties of light. A diffraction grating is nothing more than a piece of glass with scratches at regular intervals. As light passes through this grating, a truly marvelous thing occurs. White light is separated into a multitude of hues, which differ depending on the source of the light. Look through the diffraction grating at the white light coming from the fluorescent tubes so common in office buildings, and it is suddenly transformed into discrete and vivid lines of blue, green, and red—not a rainbow, where one color blends into another, but bands of sharp colors separated by blackness. Turn the grating toward a neon sign, and again you see bands of intensely vivid colors, different from those of the fluorescent light. The light from a mercury vapor street lamp produces yet different discrete bands of color. But turn the diffraction grating on an incandescent light—a tungsten filament bulb—and a continuous, rainbowlike spectrum is observed. Red blends with yellow, yellow with green, green with blue, and so on.

A truly profound question occurred to one of the students in the class. "Why does an incandescent bulb emit a continuous rather than a discrete spectrum?" Our chemistry teacher, Mr. Mullinix, pondered the question for a few moments and then confessed that he did not know. At that moment the physics teacher, Mr. Hoffman, was passing the classroom door. Mr. Mullinix pulled him into the room and had the student repeat the question. "Why does an incandescent bulb emit a continuous rather than a discrete spectrum?" This time Mr. Hoffman pondered the question for a few moments but, unlike Mr. Mullinix, came up with an answer. "Because that's the way God made it."

Two things struck me at that instant. First, Mr. Hoffman was right. While one might question the theological nature of the answer, natural laws clearly require the spectrum of an incandescent bulb to be continuous. Accordingly, Mr. Hoffman could just as well have said, "Because natural laws require it to be so." It would have been the same correct answer. The second thing that struck me was that, though correct, it was not a very satisfying answer. It was only after gaining a little background in quantum mechanics that I discovered a more satisfying explanation.

Basically, light is emitted as an electron loses energy. Once again this is a consequence of the First Law. An electron can only change energy by transferring it to something else, in this case to light. Thus the approach to making light is straightforward. Add energy to the electrons moving about molecules by hitting them with other electrons, that is, with an electric current. As these "excited" molecular electrons lose energy, light is emitted. However, the laws of quantum mechanics do not permit the electron to surrender its energy continuously but in quanta. Think about an excited electron as having been knocked to the top of a staircase. It loses energy by falling down the stairs, and in the process gives off light. The color of the light given off is related to the height of the steps. If an electron falls down a high step, out comes blue light. A low one, and out comes red light. An electron falling down a very high step will emit energetic light that cannot be seen, such as X rays, while an electron falling down a very low step will emit low-energy light that similarly cannot be seen. Such light is called infrared or microwave.

Nothing requires the excited electron to fall down the staircase one step at a time. It may fall down one, two, three, or more steps, emitting a different color light in each case. This is the origin of the spectra we observed with our diffraction gratings. Different molecules have different staircases. However, there is a general relationship dictating that the approximate height of a step in this staircase is proportional to the inverse of the size of the molecule in which the electron is excited. That is, a very big molecule will have little steps, while a very small molecule will have big steps.

So here is the answer. In a tungsten filament bulb, the exited electrons can be anywhere in the filament. It is really one huge molecule an inch or so long. In a fluorescent tube, excited electrons are contained in a molecule that's a ten-millionth this size. The steps in the incandescent bulb are tiny compared to those of all the other light sources we observed. For all practical purposes, the energy levels of the tungsten filament look more like an inclined plane than a staircase. An electron bouncing down this plane can make transitions of any energy and hence emit a continuous spectrum.

What distinguishes Mr. Hoffman's answer from mine? Both ascribe to natural law the origin of the phenomenon. In Mr. Hoffman's case it is God; in mine it is quantum mechanics. However, in my answer one finds the ability to make predictions. Whether the filament of an incandescent bulb is made from tungsten, palladium, carbon, or iron, it will still generate a continuous spectrum. Additionally, one will never find a small molecule that will produce anything but a discrete spectrum.

At the risk of being overly general, in my opinion a good deal of computational physics is concerned with determining what it is that nature allows, but not with providing predictive capabilities. To fully appreciate how and why this is done, we must take a brief detour and examine the most mind-boggling of all nature's laws, the Second Law of Thermodynamics. Like its cousin, the First Law, it can be referred to by its nickname, the Second Law.

The Second Law determines what is and what is not possible. Why is it that when something hot is in contact with something cold, the cold thing heats up and the hot thing cools down? The Second Law allows it. Why is it that heat engines aren't 100-percent efficient? The Second Law forbids it. Why is it that we remember the past? The Second Law allows it. Why is it we can't know the future? The Second Law forbids it. (For an elegant explanation of that, read Stephen Hawking's book *A Brief History of Time.*)

The Second Law asserts that something can happen only if in the process there is an increase in the entropy of the universe. Commonly, entropy is associated with disorder. Hence the Second Law is represented as constraining all possible processes to those that increase disorder. Invariably, to drive the point home, an analogy is drawn with an ordered bedroom that spontaneously becomes disordered over time—socks thrown hither and yon, bed unmade, and so forth. Unfortunately this analogy serves only to obscure the fundamental essence of both entropy and disorder.

Rather than abandon the neat (ordered) to messy (disordered) bedroom analogy, let's just refine it a bit. We will start by cataloging everything in the room—one mattress, one top sheet, two pillows, twelve pairs of socks, and so on—and then move it all out in the hall. Next, pick randomly from the list of contents, say the left shoe from a pair of brown wingtips, and then randomly select a place to put that shoe in the room. It must be in a place allowed by the laws of physics. You cannot put it on the ceiling. Continue with this process until all items have been returned to the room. Take a picture, label the picture Room 1, and then remove everything and start over. Do this until every possible arrangement of the room's contents has been exhausted. You should now have a *huge* number of photographs, and there should be no arrangement for which you cannot produce a corresponding picture.

By the way, if you do undertake this experiment, please notify me of the film you will be using. I can then buy stock in the appropriate company (or hard drive manufacturer, in the event you opt for digital photography).

Next, sort the photos into two stacks, one where all the rooms are neat and ordered and another containing the photos of all the messy, disordered rooms. As you might imagine, the "messy" stack will be many, many times larger than the "neat" stack. In actuality, we could discard the neat stack and not even notice that pictures had been lost. *This* is the essence of disorder and entropy.

Entropy is related to the number of arrangements corresponding to a system's state. In this example, the system is the room and it has two states, messy and neat. The entropy of these two states is related to the number of arrangements giving rise to each. If an earthquake were to violently shake the room, it must end up in one of the arrangements captured in the photos. Each one of these is equally likely; a specific neat arrangement is as likely as a specific messy arrangement. However, because there are so many more messy arrangements, it is much more likely that the room will end up in one of those. Thus the Second Law predicts that in an earthquake, a room in a neat state will become messy simply because this is more probable.

The previous example deals with what is called *configurational entropy.* It is related to the number of ways the constituents of a system can be arranged. This is only one piece of a system's total entropy, and, for our purposes, one of its less important pieces. A more significant part derives from the way energy is distributed among the system's components.

Say there is a system of 100 stationary balls. To this system a unit of energy is added by starting the balls moving. There are many ways the unit of energy can be added. For example, all of the energy can be given to a single ball. Because there are 100 balls, there are 100 ways to add one unit of energy to the system. The energy could also be inserted into the system through two balls, each having half a unit of energy. There are 100 ways to pick the first of these two balls and ninety-nine ways to pick the second, giving $100 \times 99 = 9{,}900$ ways to distribute energy in the system (actually it is $9{,}900 \div 2 = 4{,}950$ ways because the balls are indistinguishable, but this is a minor point). For three balls, each with one-third of a unit of energy, there are 161,700 ways to distribute the energy, for four balls 3,921,225 ways, for five balls 75,287,520 ways, and so on. You can see that the numbers are getting big real fast. If one unit of energy is evenly distributed among fifty balls, there will be a whopping 100,891,344,545,564,193,334,812,497,256 $\sim 10^{29}$ ways to distribute the energy.

At this point some may have jumped ahead and concluded that the greatest number of ways to distribute one unit of energy to the system is by dividing it equally among all 100 balls. But there are not billions, or millions, or even hundreds of ways to do this, there is only one. Each ball must have $1/100$ of the energy.

While we can easily enumerate the number of ways energy can be evenly distributed between a set of moving balls and stationary balls, we have not yet considered the number of ways energy can be *unevenly* distributed. There is no reason to assume that every moving ball should have the same energy. One ball could be moving slowly and another fast. All that is required is that the sum of the energy from each ball totals one unit. It turns out there is a specific distribution of energies among the balls that can be realized in many, many more ways than all others combined. It is called the Maxwell-Boltzman distribution, after two of the pioneers responsible for the development of statistical mechanics.

As in the case of the messy bedroom, the entropy for a system of moving balls is related to the number of ways the state of the system can be realized. Assuming for the moment that the ball experiment is conducted in a box where energy is not lost through the walls, its state is given by its temperature. Temperature is a measure of the average energy of the box's particles, which in this example is always .01 of an energy unit, regardless of the energy distribution.

Beginning from any initial energy distribution, the Second Law allows a system at constant temperature to spontaneously adopt the Maxwell-Boltzman distribution, because entropy increases in this process. If we started a single ball of the hundred moving, so that it bumps into other balls, starting them moving, in a short time the energy of all the balls will conform to the Maxwell-Boltzman distribution. At this point the energy distribution ceases to change, as any deviation requires a decrease in entropy and is not allowed by the Second Law. (In actuality, there are short-lived violations of the Second Law, but this takes us too far afield from our present discussions. For our purposes we can treat the Second Law as inviolable.)

The First Law tells us that energy is conserved and is found in only two forms, heat and work. The Second Law tells us how this energy may be distributed through the universe. It also restricts the things that can happen to those that increase disorder, where disorder is associated with the number of ways a given state can be realized.

There is a common misconception concerning the Second Law. Some suggest that it prohibits anything from becoming more ordered. This is one of the objections often raised to dispute the theory of evolution, asserting that evolution violates the Second Law, as it requires the living world to become more ordered. On the contrary, there are countless examples in which things are made more ordered. Obviously straightening up a bedroom after an earthquake is something that can happen, though perhaps not very frequently. Here is a system going from disorder to order. This is not a violation of the Second Law, though, since this law requires the total entropy of the universe to increase, but not the entropy of all of its parts.

Scientists think of the universe as divided into two parts. One of these is the system under investigation; the other is all the rest. The rest is called the surroundings. For the bedroom, the room is the system and everything else is the surroundings.

Starting with a messy, disordered, high-probability state, the room can indeed be straightened up, transforming it into a neat, ordered, low-probability state. All children know this is possible, for at the insistence of their parents, they have wasted some portion of a weekend afternoon engaged in such an endeavor. While toiling to make the room conform to the pages of *Good Housekeeping,* however, they sweat—a clear indication that work is being done and heat produced. As the heat and evaporated sweat appear in the surroundings, it becomes more disordered, as there are now more ways the state of the surroundings can be realized. The entropy of the surroundings has increased. Therefore, the decrease in the room's entropy is possible only if there is an increase in the entropy of the surroundings. The Second Law tells us that this entropy gain in the surroundings must be greater than the entropy loss in the bedroom. Otherwise the total entropy of the universe has not increased and the Second Law does not allow the tidying-up. (Wouldn't that be a *great* excuse? "Mom, I can't clean my room, it's against the Second Law of Thermodynamics.")

The point is, the change in the entropy of the system and the surroundings determines whether or not something can happen. If the number of ways a system is realized before and after some process is known, then we can calculate the system's change in entropy. If we know how much heat is lost to, or gained from, the surroundings through this process, then we can calculate the entropy change of the surroundings. Adding the two together tells us if a process can occur.

I am about to present a complex issue in extremely simplified terms, though I believe it to be appropriate. Most of the investigators responding to the "call for embrittlement proposals" thought first of the Second Law. They proposed the development of computer codes that could be used to calculate the total change in entropy associated with fracture. The objective was to answer questions like "Why do sulfur atoms embrittle steel?" by showing that the fracture of steel containing sulfur atoms resulted in a greater increase in total entropy than did the fracture of pure steel. The answer to the embrittlement question would be "Because the Second Law favors it." Of course we already knew this.

It is not surprising that the new discipline of materials design would become intertwined with the calculation of entropy changes and its component parts. After all, the goal of many computational scientists is to re-create the world on a computer. The laws of nature drive the world. So building models consistent with nature's laws is considered a good thing, and how better to build consistent models than to calculate entropy changes? Those more interested in using computers to precisely calculate energy and entropy than in designing materials had hijacked materials design.

To be fair, important new capabilities will follow the development of computer codes allowing for the calculation of thermodynamic quantities. One then could ask if vanadium, for example, would embrittle an alloy of niobium and aluminum. Into the computer goes the question, crunch, crunch, and out comes the total change in entropy accompanying brittle fracture. If there is a large increase in universal entropy resulting from the addition of virtual vanadium to the model, the answer is "Yes, vanadium will embrittle the alloy." But this is not materials design. Rather, computational trial and error has replaced its experimental counterpart. There is an advantage, if the computation can be done more reliably or faster than experiment, but it is still trial and error.

Regardless of the specifics called for in the CFPs, the intent was to develop materials design capabilities. For a materials designer, there are things more important than thermodynamics. To a designer, it is all about structure.

For almost everyone, the word "structure" evokes a strong visual. For most it is the image of something that has been built—a bridge, a building, or even an entire skyline. For a few, however, the word is, not unlike beauty, in the eye of the beholder. When asked to describe the "structure" of the Golden Gate Bridge, a civil engineer will often respond by describing it as a suspension bridge. On the other hand, an architect is as likely to emphasize its art-deco design and graceful silhouette. A traffic engineer might first call attention to the reversible lanes and one-way toll. For a metallurgist, not too far down the list of structural attributes comes a description of the main suspension cable made from thousands of laced wires and the arrangement of the individual metallic grains within each of these.

The dictionary defines structure as a building, bridge, framework, or other object that has been put together from many different parts. It is this latter half of the definition—put together from many different parts—that accounts for the egocentric interpretation of the word. For an engineer or a designer, structure becomes a personal thing: something made of many different parts that *I* can put together. What distinguishes a particular engineering or design discipline from another is only the palette of things to be put together. A civil engineer fashions designs from a palette of I-beams, reinforcing rods, and concrete. A metallurgist crafts a metallic mosaic from a palette of crystalline grains of varying shapes and composition. A chemist creates molecules with the elements of the periodic table. So to each, the concept of structure becomes intimately entangled with the arrangement of those things they are trained to put together—I-beams and concrete, metallic grains of different shapes and composition, or atoms of different types.

Yet simply putting things together from the appropriate palette does not qualify one as an engineer, or the process of putting them together as design. Design requires that the yet-to-be-made structure be characterized by predictable properties. It is not sufficient to assemble a structure, measure its properties, and then conclude that you have designed something. You must know how the structure will behave before it has been constructed. The civil engineers designing the Golden Gate Bridge knew that it would carry the weight of all the cars and trucks driving across its length. The traffic engineers knew how many vehicles could move across this bridge safely. The metallurgical engineers knew from which alloys to build the main suspension cable so that it would not sag excessively over time. In each case, this knowledge derived from well-established relationships between structure and property. These relationships are the foundations for all forms of design.

The load-carrying capacity of a suitably buttressed Roman arch provides an example of a structure/property relationship. Here the structure, a Roman arch, is characterized by the property of supporting tremendous loads. Knowledge of this relationship allowed medieval architects to design Gothic cathedrals. Before its discovery, it was impossible to construct buildings with large, open areas uninterrupted by internal columns. Thus advancements in architectural design hinged on the discovery of structure/property relationships.

A more modern example is afforded by the relationship between turbulence and drag. Drag is a property associated with objects moving through air or water. As drag increases, the energy required to propel the object at the same velocity also increases. Swimmers, runners, bikers, and skiers all seek to minimize drag in an attempt to extract the greatest possible speed from their physical effort. Typically, these athletes spend hours in wind tunnels, not making direct measurements of drag, but rather observing turbulence flow. Smoke tracers are released in the tunnel. As these tracers move around the athlete's body, turbulence, characterized by vortices outlined by the tracer, may develop. These vortices indicate an unacceptable amount of drag. Downhill skiers, for example, will then subtly shift body position looking for the perfect tuck where all vortices disappear. It is through the relationships between vortices (structure) and drag (property) that athletes design their body positions to achieve optimal performance.

Today, computational fluid dynamics permits calculation of both drag and the structure of the flow field around a moving body. For many applications, computational fluid dynamics has eliminated the need to perform wind-tunnel experiments on scaled prototypes. Though drag may be calculated directly, its value does not give a designer insight into its causes and so is simply a by-product of the calculations. As in the wind tunnel, it is the sources of vortices and turbulence the designer wishes to identify and remove. Once again, the engineer makes use of relationships between structure and properties. Turbulence is a structure that gives rise to increased drag, a property. The designer does not minimize drag directly. Instead, structure is manipulated to produce desired properties.

Getting back to the saga of materials design: The funding agencies had called for proposals to explain why elements such as sulfur embrittled steel, reducing its toughness, while other elements, like boron, transformed brittle alloys into ductile ones. All in all, these phenomena were mysterious, having no satisfactory explanation. In retrospect it is understandable that answering these questions became synonymous with the development of computational capabilities to precisely determine thermodynamic quantities. And this is exactly what happened: Computer codes became more and more accurate—so accurate that a significant mile marker was passed early in the 1990s when it became possible to calculate some thermodynamic quantities more accurately than they could be measured. One of these was interfacial energy.

Interfacial energy is one of the significant components of universal entropy change accompanying fracture. Recall that universal entropy change is made up of two parts: the entropy change of the system and that of the surroundings. In turn, the entropy change of the surroundings is controlled by the amount of heat released or absorbed from the system as a result of the process in question. So let's imagine something that is going to break into two parts without any accompanying dislocation motion. The part on the left side we will call *E*, and the part on the right we will call *F*. When joined together it will be represented as *E–F*. The process of breaking it can then be represented symbolically as $E–F \rightarrow E + F$. The energy necessary to make this process go is the difference in energy of *E–F* and the energy of $E + F$. This is figuratively the energy of the "–" holding *E* and *F* together. The greater this energy, which must be supplied from the surroundings, the more the entropy of the surroundings must decrease. By the Second Law, *E–F* will fracture along the interface only if there is an increase in entropy to offset its decrease in the surroundings. For obvious reasons, calculating interfacial energies was seen as an essential capability of computer codes needed to explain why things break.

Interfacial energy became the hot commodity. The better a computational scientist could calculate it, the more papers he (no sexism intended; everyone I know working in this field is a man) could publish. The more papers published, the happier the program monitor; the happier the program monitor, the greater the likelihood of getting more funding. With more funding, the scientist could write better code to calculate interfacial energy even more accurately and publish more papers, et cetera, et cetera. The key here is for the scientist to show that his calculation is more accurate than everyone else's. This demonstration required something for comparison, not another calculation, but a good solid experimental measurement of interfacial energy. Unfortunately, measuring interfacial energy is extremely difficult and not especially accurate, particularly in metals and alloys, where, even if brittle, some of the work of fracture goes into making dislocations. However, for some specialized interfaces there are indirect methods that yield fairly accurate results. These special interfaces, called stacking faults, became the benchmarks through which modelers could validate their codes.

Almost overnight, every modeler worth his salt was busy calculating stacking-fault energies. And within a very few years the variation in calculated stacking-fault energies between researchers was smaller than the experimental error of the measurements. At this point it was hard to substantiate claims as to which code was the best. It was not unlike arguing about the best brand of pickup truck. Is it Chevy, Ford, or Dodge? Everyone knows it's the Ford F-series. Of course, that's what I drive when I am not zipping around in my Honda S2000, which is the best sports car ever made.

Nonetheless, having spent millions of dollars to develop and perfect computer codes, which determined interfacial energy more accurately than measurement, the researchers turned their tools to the real task—embrittlement.

The embrittlement problem now appeared straightforward. Sulfur atoms embrittle iron when segregated to grain boundaries. So calculate the interfacial energy of a pure iron grain boundary. Next calculate the interfacial energy of the same boundary, only this time sprinkle a few sulfur atoms around. The results of these calculations showed that sulfur atoms decreased the interfacial energy of iron grain boundaries. The problem had been "solved"; calculated results were consistent with the Second Law.

Fast and furious followed further calculations. Now we knew why boron atoms toughened a brittle alloy of nickel and aluminum. In accordance with the Second Law, these atoms increased the interfacial energy of alloy grain boundaries—and phosphorus embrittled both nickel and iron through a decrease in interfacial energy.

Oh, happy days, computation had triumphed; materials design was just around the corner. That may have been what the proud program monitors, sponsoring all this research, thought. Armed with these truly impressive computational capabilities, the funding agencies now actively encouraged collaboration between materials designers and computational scientists. The program monitors wanted to see some materials design, and were anxious to sponsor research to that end. Metallurgists and materials scientists teamed with computational physicists to make tough new high-temperature alloys. The "design"-oriented members of these teams asked just one thing of their "computationally" oriented colleagues, "Show us the structure." That is, show us the structure responsible for brittle fracture.

The demand for structural information presented itself as perfectly reasonable, and at first it was a demand that was thought to be easily satisfied. After all, with the incredibly accurate calculations of interfacial energy came incredibly accurate calculations of interfacial charge density—that is, where all those moving electrons were located. In turn, the locations of the electrons shaped the bonds. So a quick comparison of the computationally determined charge density of brittle and ductile materials would give the design folks just what they wanted. Or so it seemed.

The papers relating charge density to fracture properties began to appear in the early 1990s, and mysteriously there was no consensus. Calculations that had given almost identical values for interfacial energy and nearly indistinguishable charge densities were interpreted as giving rise to very different bonding. One researcher would claim that the brittle properties of nickel aluminide (an alloy of nickel and aluminum where the ratio of nickel to aluminum atoms is one to one) were due to "directional bonding." Another researcher would come to the opposite conclusion, attributing nickel aluminide's brittle nature to reduced directional bonding. Still others made systematic comparisons across an entire series of alloys, concluding that there was no feature of the charge density that could be associated with brittle or ductile behavior. A few of the computational physicists who had contributed the most to the development of highly accurate codes now espoused the belief that brittle behavior was not controlled by the electronic structure at all.

Around 1995, CFPs to integrate theory with materials development specifically included language to the effect that materials design proposals would not be considered. It appeared as if the brief but expensive materials design programs had ended in failure.

INSIDE MATERIALS BY DESIGN

CHAPTER 11

While the computational people struggled to find relevant structure in the charge density, I wrestled with the problem of comparing those same densities. At the time, no one recognized that we all were confronted with one and the same challenge.

I received my Ph.D. in May of 1983. Like most spring graduations, the ceremonies were held outdoors on the Great Court, a spectacular setting overlooking the Charles River and the Boston skyline. Since New England's adverse weather had played such an important part in my graduate education—motivating my choice of graduate school as well as providing research insight—it seemed somehow appropriate that the day should be marked by rain. Not a heavy rain, which would have caused the observance to be moved inside. Rather it was a drizzle, which, halfway through the three-hour ritual, had drenched all the graduates. This provided MIT's president with the opportunity to make puns, comparing the soaking the graduates' families were now experiencing to the one they had taken over the last four years. Despite the rain and the president's bad puns, I remember the day as one of my life's highlights.

In the year preceding graduation, I began looking for a postdoctoral position. Greg Olson and Ron Latanision offered me the opportunity to continue my work at MIT. But I also wanted to explore other options, particularly those in parts of the world close to better skiing. High on my list of desirable locales was Los Alamos National Laboratory (LANL) in New Mexico, where I applied for a prestigious postdoctoral appointment. In January, John Wood, a staff scientist from the lab, contacted me while I was in Aspen on a ski trip. Evidently he had reached me by first calling his friend Keith. I had made the short list and was one of about half a dozen candidates who would be interviewed for the position. Several days later, Cheryl and I were in New Mexico. I spent two days interviewing at the lab while Cheryl toured Los Alamos and Santa Fe. We both fell in love with northern New Mexico.

To my chagrin, I was not offered the postdoctoral position hoped for. However, one of the scientists with whom I had interviewed, Pat Martin, arranged for me to spend protracted periods at the lab. This worked out just fine. I accepted Greg and Ron's offer with the understanding that I would be spending long periods in Los Alamos. For the next year and a half I commuted between Los Alamos and MIT. In 1985, my postdoctoral travels ended when I was appointed a staff scientist at Los Alamos National Laboratory.

As an aside, it seemed we could not leave New England without one last kick from the Boston weather. The day before the moving truck was scheduled to pick up our furniture, the moving company's packers put everything in boxes and stacked them in the front room, available for quick loading into the truck. We had planned to spend one night in the house, sleeping on the floor, and then begin the drive to Los Alamos after the moving trucks were packed and on the way. The day dawned, however, with a prediction that a hurricane would slam into the New England coast within thirty-six hours. The exact location of landfall was unknown. All day the wind gusts mounted, and by midafternoon we received a call from the moving company—they were not coming. Our house was quite close to the coast, and the weather advisories were warning us to tape and cover windows and make other emergency arrangements. Of course, everything we owned was neatly packed into boxes, and it would have been impossible to find anything. So we rode the hurricane out. It was our first, and not particularly severe; the winds never exceeded eighty miles per hour. Yet a hurricane closing the quotations on our time in New England still seems remarkable.

At Los Alamos I would be investigating a new aspect of

fracture. While Greg and Ron were concerned with embrittlement, particularly embrittlement of iron and nickel alloys, the Los Alamos scientists were exploring the opposite phenomenon, cohesive enhancement. Just as there are elements that decrease a metal's toughness, there are others, called cohesive enhancers, that have the opposite effect, increasing toughness. For example, a little thorium will increase the toughness of iridium. Remember that iridium is the metal used to encapsulate plutonium-238 in radioisotope thermoelectric generators. Though a tough metal in its pure form, iridium will fracture along grain boundaries. But add a pinch of thorium and the resultant alloy is incredibly tough. What is known is that when added to the iridium, the thorium atoms move about until they reach grain boundaries, where they localize. This process is called grain-boundary segregation and the thorium atoms are said to be grain-boundary segregants. What is unknown is how the thorium atoms act to inhibit grain-boundary fracture. Similarly, boron atoms segregate to the grain boundaries of many metallic systems, and this segregation is frequently associated with improved toughness. So common is the improvement in toughness due to the addition of boron that it is sometimes called "the boron effect."

Los Alamos's interest in cohesive enhancement, and specifically the boron effect, was motivated by a Department of Energy program manager. Bear in mind that this was the period immediately following the first publications by Bryant and Messmer, and funding agencies were running around trying to get a piece of the action. What better place for a DOE program manager to go looking for action than

Los Alamos? After all, the lab is one of DOE's weapons labs, and hence is supported primarily through DOE funding. Additionally, the largest, fastest computers in the world were located at LANL, along with some of the world's best computational physicists. So a DOE program manager goes to Los Alamos and says, "Here's the money, explain the boron effect." The lab accepts the money and a new program is born. The program manager christens it "Materials by Design."

The name gives us a clue that the program manager thought of the boron effect as a stepping-stone to materials design. I'm sure he anticipated that his funding would seed the development of a whole suite of materials design software. As part of my research responsibilities, I was assigned to the Materials by Design program. I was young and naïve, and thought the objective of the program was to explain the boron effect, which is what I set about to do.

Once again the problem required comparing the charge densities of a fractured to a stretched material. I pictured the electrons zipping around at a metallic grain boundary. These electrons, like glue, literally hold the two sides of the boundary together. As a load pulls the grains apart, the electrons at the interface can do one of two things: either move into the broadening interface or move closer to the forming surfaces. In moving closer to the forming surfaces, the charge density more closely resembles that of two noninteracting grains. This must be what happens during fracture: The electrons move toward the surface. In contrast, electrons moving into the broadening interface are resisting fracture. I thought, whether an impurity atom acts to embrittle or enhance cohesion will be mediated by how the atom's electrons move in response to the load. If its electrons redistribute toward the surface, fracture is being promoted and the element is an embrittler. In the alternative, if its electrons move into the broadening interface, it is a cohesive enhancer.

The principles that govern the redistribution of an atom's electrons as nearby atoms move are well understood and in some situations are predictable. The charge density around an isolated atom of any element is spherical. That is, if you were sitting on the nucleus of any isolated (far removed from all other atoms) atom, the charge density would look the same in every direction. But bring other atoms nearby and the electrons shift, changing the shape of the charge density. There are many possible shapes for this shifted charge density, but we will be concerned with the two principal ones.

Start with a sphere—say a ball of clay on a table. Press down on the clay as if to form a pancake. This resulting shape is called an oblate spheroid, characterized by an equatorial diameter greater than its length from pole to pole. Re-form the sphere of clay, but this time roll it between your hands a couple of times as if you were starting to sculpt a snake. The resulting shape is called a prolate spheroid and is longer from pole to pole than across the equator.

A simple formalism called crystal field theory, which was developed in the 1930s and 1940s, predicts which of these three shapes—spherical, oblate, or prolate—an atom prefers. As always, we rely on the laws of thermodynamics to make this assessment. Starting with a spherical shape, we compute the change in the entropy of the surroundings that would result from a shift to oblate or to prolate. Generally, at most one of these shifts will result in an increase in the entropy of the surroundings, indicating which charge density shape is generally preferred by atoms of this element.

Remarkably, atoms like boron and thorium, which prefer the prolate charge density, act as cohesive enhancers, while those preferring the oblate shape, like sulfur and phosphorus, act as embrittlers. The mechanism is obvious: A prolate-preferring atom moves charge density into the broadening interface region, effectively becoming more prolate and resisting fracture. At the other extreme, an oblate-preferring atom becomes more oblate as grains separate, moving charge into the developing surfaces. This promotes fracture.

As far as I was concerned, the boron effect had been explained, and without doing a single calculation or experiment. I had relied only upon the fundamentally sound and well-accepted foundations of crystal field theory. In marked contrast to the approval I received from the material, scientists at MIT, my computational colleagues at Los Alamos were neither impressed by nor supportive of the explanation. As I had never worked with a group of hard-core computational scientists before, I assumed that this attitude was par for the course. Acceptance would come only after I worked out all the bugs and published the work in a peer-reviewed journal.

My goal became to publish this theory of embrittlement and cohesive enhancement in *Physical Review Letters*—

a very well-respected journal. As I had previously submitted manuscripts to *PRL* without success, it seemed expedient to collaborate with someone who could bring a new perspective to the problem. Dimitri Vvedensky was just the guy.

Dimitri was another of Keith Johnson's graduate students who was finishing up at MIT just I was beginning. In the intervening years he had moved to London, where he was an instructor at Imperial College—England's MIT. I had always been impressed by Dimitri, and thought this would be a great opportunity to work with a world-class scientist. My supervisors at the lab came up with a little money to support Dimitri's visit. It was done. He was on his way.

Dimitri and I worked well together. We used the Slater and Johnson code to show that cohesive enhancers and embrittling elements behaved exactly as expected. The charge density of boron atoms segregated to grain boundaries was prolate. Similar calculations indicated that at a grain boundary the charge density of the embrittling element sulfur was oblate. We summarized the results of our calculations, along with a general theory of the boron effect, and published the whole works in *Physical Review Letters.*

Surely now my Los Alamos colleagues would embrace the model. But no, it seemed there were other troubles. The Slater and Johnson code was, admittedly, quite old and inaccurate; it could not be used to calculate thermodynamic quantities such as interfacial energies. Consequently, since these were not state-of-the-art calculations, those who were developing such computational tools largely ignored our results.

Back at Imperial College, Dimitri and his graduate students were hard at work writing a quantum-mechanically based computer code that was different than all others. What made this code different was advertised in its name, the layer-Korringa-Kohn-Rostocker (*LKKR*) method. Other codes of the day used a super-cell approach to approximate the charge density or energy of an isolated interface. The "layer" part of the LKKR code circumvented this approximation and was used to give us the first views of the charge density at isolated interfaces. But not just charge densities. Like so many others, we also wished to validate the code through the calculation of thermodynamic quantities such as stacking-fault energies.

Part of the development of the LKKR code took place at Los Alamos. James MacLaren, who was a student at Imperial College working with Dimitri, accepted a postdoctoral position at the lab. Starting in 1988, James put the finishing touches on the LKKR code, and in no time at all we were in business studying grain boundaries.

The calculations using the Slater and Johnson code were repeated, this time using the LKKR code. These calculations confirmed the relationships between a prolate charge density and cohesive enhancement and an oblate charge density and embrittlement. Still, from the scientists at Los Alamos working on the boron effect, there was nothing but silence. It would be years before I finally understood what had happened. The purpose of the Materials by Design program was to develop software that would reveal the cause of the boron effect. That the effect could be rationalized with a fifty-year-old theory was antithetical to the Materials by Design concept. It did not matter that we were using the most advanced codes to confirm our theories; the codes were ancillary to the discovery.

Significantly, several other papers have been published in prominent journals purporting to have uncovered the cause of the boron effect. In every case, boron's cohesive enhancement has been attributed to its prolate charge density. Occasionally the authors of these studies have cited the *PRL* article by Dimitri and me. More often, however, even when well aware of our earlier paper, they attribute their insights to the use of more accurate codes.

All things considered, the environment in Los Alamos was not supportive. It seemed that I had misunderstood the whole reason for my being there. So I began looking for a more encouraging environment. In January 1990, I began to relocate to the Colorado School of Mines, fifteen miles from where I had grown up and fifteen miles closer to the great ski areas west of Denver.

With the 1990s came a waning in support for materials design research. A few of the more visible programs, like those at Los Alamos, became serious modeling efforts. These often pushed the limits of computation, but seldom shed light on the structure or properties of real materials. However, there was an upside. Just as in nature, where mass extinctions make room for new species, a couple of innovative materials development programs were able to take hold.

These emerging programs differed from the now-vanished efforts of a decade earlier in that they were directed by metallurgists who thought of materials as complex systems governed by structure/property relationships.

One of these was run by none other than Greg Olson, who had relocated to Northwestern University in Evanston, Illinois. Yet another program, directed by Dennis Dimiduk, evolved at the materials laboratory at Wright-Patterson Air Force Base in Dayton, Ohio.

Significantly, for me, both Greg and Dennis felt that while fracture was an important aspect of materials design, it should not be its centerpiece. This took tremendous pressure off those few of us who were still studying fracture. The expectation that we would discover some amazing "pixie dust" that would endow an ordinary material with incredible toughness was removed. Their only expectation was that we would discover the structure/property relationships governing fracture. And it was to that task that I would devote the next ten years.

MATERIALS BY DESIGN: RESURRECTION

CHAPTER 12

Though one aspect of the fracture problem had been cracked, we were still far from finding the complete answer to why things break. Comparing the charge densities of grain boundaries containing boron atoms with those containing sulfur atoms was like comparing apples and oranges. There was just no question that these were very different kinds of atoms, producing very different effects. When it came to more-subtle property variations, the charge density seemed to give no clue. For example, one of the systems of interest to my collaborators at Wright-Patterson is a group of alloys known as B2 aluminides. "B2" is shorthand for the alloy's crystal structure. And though B2 nickel aluminide has a number of attractive high-temperature properties, it is brittle—not as brittle as B2 cobalt aluminide, but more brittle than B2 iron aluminide. The interest was in explaining the differences in mechanical properties as well as finding the element or elements that when added to B2 nickel aluminide would make it tougher. In the absence of understanding, the search for these elements proceeded by trial and error.

As far as computational studies were concerned, calculated interfacial energies were consistent with the observed trend in toughness, yet comparing the charge densities of these three alloys was like comparing nearly identical triplets. There were slight differences, but none seemed to correlate with observed properties. It made no sense at all. Accurate calculations of interfacial energy must simultaneously generate accurate charge densities; one requires the other. It was as if we could calculate the effect, but the cause remained hidden in plain sight.

I guessed that the problem was rooted in the way we thought about the chemical bond—picturing the bond too qualitatively to permit a detailed comparison of charge densities. A new understanding of the relationship between charge density and bonding was required.

Conventional models of the chemical bond had not changed much in nearly seventy-five years. We have gotten better at calculating charge densities, but we still think of the bond in terms of the amount of electronic charge located between atoms, a concept that falls short of the reality. As atoms move apart, charge stretches thin but never vanishes. Thus the connection between atoms is never broken and the concept fails to describe fracture. Some other feature of the charge density must better represent a bond and how it breaks. Chemist Richard F. W. Bader of McMaster University discovered exactly what this feature is: topology. Bader was the first to describe the chemical bond in terms of the charge density's topology.

Topology is a branch of mathematics that describes the nature of connections that endure within an object as it is stretched or squeezed. (Cutting or combining shapes destroys an object's original topology.) For example, a teacup and a doughnut are topologically identical. If you use the hole in a clay doughnut to form a handle, you can sculpt the doughnut into a teacup without cutting the clay. On the other hand, a ball of clay cannot be molded into either a doughnut or a teacup without punching a hole in it and separating parts that had previously been connected.

Bader used his revolutionary idea about chemical bonds to analyze organic molecules, but I found in his observations a fascinating way to look at a metallic solid. Fracture is a process that changes the nature of the connections within an object, so it made sense to describe it using topology.

A topological connection between atoms can best be explained by comparing charge density to a mountain range. In this picture the altitude corresponds to the density of the charge: Peaks represent areas of greatest charge, basins are points of least charge, and so on. Two atoms are topologically connected if there is a ridgeline of charge density between these peaks. Such ridgelines correspond to the conventional picture of chemical bonds between atoms.

Topology tells us that the existence of a ridgeline is guaranteed if there is one special point in the charge density, a saddle point. In terms of our terrain analogy, a saddle point corresponds to a mountain pass. Think about riding a bicycle up a mountain pass. Initially, you must work hard to pedal the bike uphill. Stop pedaling and you roll backwards. At some point you crest the pass and can coast down the other side. That point right at the top where you make a transition between pedaling and coasting is a special place.

In your mind's eye, stand at this special place and look around. Looking back the way you came, the terrain drops off, just as it does if you proceed ahead, down the pass. But look left of the path you rode, and the terrain is uphill toward a mountain peak. Look right, and again the terrain is uphill toward another peak. There is no other place on this pass with this property: Forward and backward the terrain falls, left and right it rises. This place is a saddle point and requires that there be a ridgeline running through it and connecting the two peaks on the left and right.

The saddle point is one kind of critical point. Critical points are simply places where you can, if sufficiently talented, balance yourself on that bicycle. You could also balance yourself at the bottom of the basin, where looking forward or backward, left or right, the terrain rises in all directions. Such a critical point is called a minimum. Finally there are the mountain peaks, where you could also balance yourself, but from here all directions are downhill. This critical point is called a maximum.

Critical points provide the minimum information required to reconstruct the shape of the terrain. If I knew only where the mountain peaks, basins, and passes were, I could reasonably describe the local landscape. I could describe it even better if I knew the elevation at each of the critical points. Then I would know how much work would be required to ride a bicycle over a given pass. Better yet would be knowledge of how curved the terrain was around a critical point. Now I would know if ropes and other paraphernalia were needed to reach some mountain peak, or if it could be scaled by a less arduous final ascent. The key here is that knowledge of the overall shape of the terrain can be had by studying the properties of critical points in greater and greater detail.

In our analogy, the minimum information needed to distinguish a critical point is its location and elevation and the shape of the terrain immediately around it. In turn, this shape can be characterized by the point's principal curvatures, which indicate how rapidly the altitude is changing in two perpendicular directions. The principal curvature is negative when the altitude decreases, as from a mountain peak. Conversely, positive curvature occurs in any direction in which altitude increases.

At the saddle point between two mountains, the altitude will increase in the direction of the nearby peaks and decrease in the perpendicular direction. The pass has both positive and negative principal curvatures. A minimum critical point is characterized by two positive principal curvatures and a maximum by two negative principal curvatures.

To describe a bond, we need to extend this picture to three dimensions in which we can move up and down as well as right and left, backward and forward. Every critical point now has three principal curvatures. For example, a saddle point with two negative and one positive principal curvature indicates the existence of a ridge of maximum charge density connecting two atoms.

Picture yourself floating in space at such a saddle point. Off to the right and left are atomic nuclei. Neither can be seen, however, as each is surrounded by a dense cloud of electrons. Move up or down, backward or forward and, as if you were descending through a cloudbank, the fog diminishes. Alternatively, move toward one of the nuclei, and the electronic fog engulfing you becomes denser. Follow the path to the nucleus along which the electronic fog is densest and you are moving along the bond.

The charge density may have three other kinds of critical points: a maximum, with three negative principal curvatures; a minimum, with three positive principal curvatures; and another kind of saddle point, with two positive and one negative principal curvatures.

Using the topological description of bonding, along with the magnitude of the principal curvatures involved, we can analyze the structure of a chemical bond more quantitatively. First, again consider a saddle point in two dimensions using the analogy of the mountain range. Starting at the pass between two mountains, you could begin walking in any of four directions, two around each peak, along which your altitude would not change. If four people started at the pass, and each took off in a different one of these directions, initially their paths as seen from above would form an X, where the mountain peaks are located to the right and left of the X.

The angle formed by the top two legs of the X tells us how wide the pass is relative to the distance between peaks. A large angle indicates a narrow, steep pass with a tightropelike ridgeline. A small angle signifies gently descending terrain away from the ridge. An angle of zero signifies the vanishing of the ridgeline. The angle thus provides information about the shape of the ridgeline.

The mountain-pass analogy is more complex when used to characterize the charge density in three dimensions. Once again, imagine floating in space at a saddle point. There are now two independent sets of four directions along which the charge density will not change. Each set will form an X in a different plane. First consider the plane containing the forward, backward, right, and left directions. To your right and left the charge density increases. Ahead of you and behind you the charge density falls away. Hence there must be directions in this plane along which there is no change in the density. As before, these directions form an X. Next consider the plane containing the up/down, and right/left directions. Again, to the right and left the charge density increases, while up and down it decreases. Describing the shape in this plane requires another X of directions in which the density does not change.

The two Xs required to picture the shape of the charge density in three dimensions need not be characterized by the same angle. The ridgeline could be very steep in the up/down direction and relatively broad in the forward/backward direction. Hence these two angles describe the charge density around a chemical bond.

That two angles could be used to characterize the chemical bond was interesting but not of much practical importance, unless there was a relationship between the angles and properties. Then a new structure/property relationship would have been uncovered.

I fully expected such a relationship would exist, as the angles were "invented" in response to my need to understand bond-breaking. I reasoned that as two sides of a material were pulled apart, the angles characterizing the bonds holding it together would begin to decrease. At some point the angles would reach zero, indicating the bonds had broken. It seemed reasonable that the smaller the angle, the more closely the charge density resembled that of the fractured material.

Alternatively, instead of pulling the material apart, the two halves could be sheared. In this case the angles of the bonds resisting shear would decrease. When these angles vanished, the bonds resisting shear would break. So it also seemed reasonable that the smaller this angle, the more closely the charge density of the native metal resembled that of the deforming substance.

From this line of analysis, it seemed that the competition between ductile and brittle behavior would boil down to comparing different angles. A ductile material would be one in which the angle that changed during shear was small compared to the changing angle during elongation.

Verifying the validity of such a structure/property relationship was straightforward. Calculate the charge density of many pure metals. Use the calculated charge density to find the angles characterizing the bonds. Compare these angles with the known elastic constants. These measure a material's resistance to elongation and shear. If the relationship were valid, a correlation would be found.

There was just one problem with the plan: Neither the Slater and Johnson code nor the LKKR code was suitable for the task. Both found the overall charge density by first finding the density in little regions and then piecing these together. The effect was much the same as building a ball with Lego building blocks. Instead of one smooth sphere, you get a very jagged-looking thing, where the corners of every block introduce false critical points.

Fortunately, one of the first things the computational guys had done when constructing highly accurate codes was do away with the piecewise assembly of the charge density. James MacLaren, who was now at Tulane University, had written one of these codes, just what was needed to calculate the angles' characterizing bonds. And sure enough, just as I had hoped, the correlation was found.

In short order, other correlations began to appear. One was found between the angles and the ductile-to-brittle transition temperature of pure metals. And that sought-for feature of the charge density, which would explain the relative toughness of B2 iron, cobalt, and nickel aluminide, was also found. Another correlation relating slip to "Gaussian curvature" was discovered. We still do not understand the basis for this correlation, but we had a good reason for seeking such a connection.

The Gaussian curvature of the charge density is the product of the three principal curvatures. By way of example, take a piece of paper as a two-dimensional illustration. The paper is flat, so its curvature in any two perpendicular directions is zero. Obviously the product of zero with zero is zero. So the Gaussian curvature of the paper is zero.

Rolling the paper into a tube presents no difficulties. Yet if we were to try to bend the tube such that the two open ends were brought together forming a doughnut, we would be unsuccessful. Why is it we can form the tube but not the doughnut? The reason is found in comparing the Gaussian curvature of the tube, the doughnut, and the flat piece of paper.

On the one hand, around the tube the paper is curved and hence is characterized by non-zero curvature. Yet along its length the tube is still flat and hence its curvature in this direction remains zero, therefore, the Gaussian curvature of the tube is zero. The doughnut is curved in both directions and must therefore have a non-zero Gaussian curvature.

A theory of geometry asserts that one shape can be turned into another without stretching only if the Gaussian curvature of the shape remains unchanged. Thus the paper, with zero Gaussian curvature, can be turned into the tube, also with zero Gaussian curvature, but not into the doughnut.

The Gaussian curvature tells us something about what is allowed when turning one shape into another and it is for this reason that I looked at this quantity for possible structure/property relationships.

Though structure/property relationships have been discovered, we are still one step away from design. Design requires the ability to change structure to produce desired properties. The structure/property relationships tell us what the charge density should look like. But how is this density to be put together? This is not as daunting a task as it appears.

Though I oversimplify a bit, every element has a specific shape to its charge in a given environment. It is as if we had hundreds of building blocks of varying shapes and sizes. From these, almost any charge density desired can be built. I used this approach to predict that substituting iron for nickel atoms in B2 nickel aluminide would improve toughness.

By the time I made this prediction, the empirical search for alloying elements to improve nickel aluminide had been under way for nearly fifteen years, so there was every reason to believe that other researchers already knew whether iron produced the desired properties. At a conference in 1998, I discovered that relevant experiments showed that 10 percent iron substituted for nickel had exactly the predicted effect.

More important than the prediction is the corollary that came along with it. Not only will iron, and perhaps manganese, additions improve those mechanical properties of nickel aluminide that are a consequence of its charge density, but these are also the best additions. There is no need to look to chrome or platinum or some other combination of elements. Trial-and-error searches can never be used to rule out untried possibilities. With trial and error, you never know when to stop.

Throughout the 1990s, I worked with Greg Olson and

Dennis Dimiduk, looking for relationships between charge density and properties. The group at Wright-Patterson has perfected some of the most advanced computational tools available. With these we can look at the structure of the charge density around a moving dislocation. We expect to find very subtle relationships between the bonding and the ease with which dislocations move. In collaboration with Greg's group, we are looking for relationships between the charge density and diffusion and solubility of oxygen in niobium—an important consideration for the next generation of high-temperature materials.

By 2000, Greg Olson's materials development program at Northwestern had blossomed into a joint industry-university initiative. As a spinoff of the research conducted at Northwestern, Greg founded a materials development company under the name Questek, and trademarked Materials by Design as a company trademark. Materials design is apparently not dead yet.

IT'S BROKE. WE'VE GOT TO FIX IT

CHAPTER 13

So there is structure to the charge density—structure that gives rise to properties and structure that can be manipulated. The problem is solved. It will be only a short time before new materials with dramatic new properties are popping up all over—well, probably not.

There is a broadly held belief that once the basic science has been done, the industrial sector will step in and do the technological development necessary to commercialize new discoveries. If a viable product does not flow from the science, it is assumed that either there was no commercial value to the discovery or still more basic research is needed to unlock its potential. Unfortunately, this belief is just plain wrong.

Let's take as a hypothetical example Greg Olson's company, Questek. Certainly the researchers there are capable of developing a specialty steel, integrating the structure/property relationships uncovered in my research into their "systems approach" to materials development. In fact they collaborated with me on a good deal of this work. Let's assume the steel is so superior that another company takes notice and decides to implement the same systems approach. The first thing the company needs is someone trained in this method. Where do they go to find such a person? There is only one place: Greg's Steel Research Group at Northwestern University.

As research groups go, SRG is large, and so it is likely that within two years a suitable candidate will graduate. Off he or she goes to set up a materials development program. First this person needs an advanced computer system, access to the software used at Northwestern and here at CSM, and researchers from the company trained in its use. Some of this code is commercial and can be purchased off the shelf—for $50,000 to $100,000—but some is what we call research grade. This is code kludged together by students interested in graduating, not in making things easier for the next user. There are no instruction manuals. Hence, along with this research-grade code one needs a graduate student or postdoc who is paid to teach an introductory class—Code Use 101—to the company's researchers.

After six months to a year the scientists and engineers are starting to understand the systems approach. Some have mastered the kludged-together quantum-mechanical codes by repeating the numerous calculations performed by previous users. They are now performing the first calculations on systems of interest to the company.

About this time the CFO wants to know what has been produced with the two million or so dollars spent in this materials development program. Sadly, there is nothing to show, other than researchers thinking about materials development in a different way. And with that, the program is shut down.

This scenario is not as fanciful as it may seem. The same initial excitement that causes program managers to throw money at materials design seems to sweep up midlevel corporate types. Though they remain great supporters of the concept, companies still have financial interests, and the costs associated with launching new technologies can be astronomical.

Despite all evidence to the contrary, the concept that technological development must proceed through the commercial sector persists. This idea simply ignores the fact that the federal government launched many if not all of the major technologies of this century. Consider just a few of them:

- Automotive technology would probably not have advanced to the point it has today if not for the Federal Highway Act of 1956, when, under President Eisenhower, the federal government committed itself to building the National System of Interstate and

Defense Highways. Financing for this $130-billion project, which was not deemed complete until 1993, came primarily from state and federal taxes on gasoline. In part it was the existence of the Interstate Highway System that produced the demand for automobiles.

- Integrated circuits represent the enabling technology for computers, satellites, and telecommunications, to name but a few. The first integrated circuits were invented in the late 1950s, but were prohibitively expensive, as far as consumer products were concerned. However, the air force didn't care about cost, since it needed reliable electronics for the control and guidance aboard missiles, especially the Minuteman intercontinental ballistic missile of the 1960s. The Air Force's demand for integrated circuits not only drove the cost of production down, but also advanced silicon-processing technologies, leading to the development of the microprocessor in 1971.

- The microprocessor is the enabling technology of personal computing, but this is only the latest development in the history of electronic computers. The first electronic computing machine, ENIAC (Electronic Numerical Integrator and Computer), was invented in 1946 at the University of Pennsylvania. The army funded the three-year development of this machine in hopes of finding a quicker way to calculate the trajectory of artillery shells. In the 1950s, the Whirlwind computer was developed at MIT's military-funded Lincoln Laboratory; its purpose was to process information from radar regarding attacking aircraft and then control intercepting aircraft or missiles. It would not be until 1964 that IBM introduced the first computer used widely for nonscientific and commercial applications, the System/360.

- Since the advent of IBM's System/360, national laboratories, the military, and research universities have maintained state-of-the-art computer systems. Under the auspices of the Advanced Research Projects Agency of the U.S. Defense Department, these computer systems were networked early on, forming what was known as ARPANET. Over this network, users could transfer files, share software, and log in remotely. In 1971, ARPANET became the core of the Internet.

If history is any indicator, before the discovery of structure within the charge density leads to new technologies and innovative approaches to the design of materials, there will be either a very long or very expensive incubation time. If it is deemed a matter of national security, user-friendly computer codes must be written and researchers in critical industries must be trained to use and interpret the results. If not, then the process of conversion will be slow—training one student at a time to think in a different way.

This book has been concerned as much with the system through which science is done as with the science of why things break. Appropriately, those two story lines converge here, because the system that promotes science and technology is much like the steel in the *Titanic*. It is defective. Of course, at the time no one knew of the *Titanic*'s little problem. Only hard-won experience would uncover the nature of the defect.

The system that promotes science and technology today was placed in service at a time when we had little experience regarding the interrelationships among governmental policies, science, and technological advancement. Though largely unlearned, the lessons of the last sixty years show how inefficient the present system is. It is extraordinarily important that this system be upgraded, as science and technology have become integral parts of every aspect of our culture. Economic health is inextricably entwined with technological innovation. National defense cannot be isolated from weapons, surveillance, and intelligence technologies. Health care, domestic security, law enforcement—all of these increasingly depend on a healthy science and technology base. And though our politicians are more than happy to present energy policies, health-care plans, and foreign and domestic economic plans, they never present a comprehensive plan involving a significant technological component.

When was the last time any president provided a plan to stimulate the economy through directed technological advances? To the best of my knowledge, never, despite the fact that economic growth has often been attributed to specific innovations, as the prosperity of the 1990s was attributed to the development of the Internet. Why is it that politicians don't seek technological solutions to our economic problems? The answer to this question can be found in one last story:

In the early 1990s, I traveled to Washington, D.C., to meet with the contract monitor of one of my research grants. Since I was in the city, I decided to look up a colleague, Michelle Donovan. Michelle had for a number of years been working in the area of scientific policy for the Department of Energy. She had done such an outstanding job with DOE that she had been loaned, so to speak, to the White House to work for the Presidential Science Advisor.

Somehow I had misplaced Michelle's telephone numbers, but I figured it would be an easy matter to find the Office of the Presidential Science Advisor in the Washington phone directory. Under the government listings, which in the Washington directory go on for pages and pages, there was no separate listing for "Science Advisor" or "Presidential Science Advisor." However, there was a rather extensive listing for various functionaries associated with the White House. Although there was a number to reach the White House groundskeepers, there was no number for the Presidential Science Advisor or an Office of Science and Technology under the White House listings. In fact, I could not find a single reference to an office connected with science or technology associated with the executive branch of the government and only one listing, as a sub-sub-category, under the legislative branch. Reasoning that I had just overlooked the listing, I called the White House switchboard and asked to be connected to the Presidential Science Advisor's office. The operator asked me to repeat my request no fewer than three times before informing me that she had never heard of such an office. At this point I simply gave up, called the DOE, from which Michelle was on leave, and got her new number from them.

My first question, after the customary exchange of pleasantries, was, Where had I gone wrong? Was I mistaken and the office where Michelle now worked was not connected to the White House? That wasn't it. Michelle assumed I must be wrong about the listing. While I followed her directions and made my way to the Office of the Presidential Science Advisor, she perused the government listings in the Washington phone book. Sure enough, she couldn't find a listing, either.

Why is it that the Presidential Science Advisor was so easily overlooked? My feeling is that he was, and probably still is, perceived differently from, say, the Secretary of State or the White House groundskeeper. Both the Secretary of State and the White House groundskeeper have some ability to shape the future. As most of the public has the perception that technology is something that "just happens" when the science is done, the Presidential Science Advisor has no such ability. He, or she, is perceived in the same light as a reporter, who does not make the news but is around to inform us when something happens. We only take note of such people when something happens.

Is it possible that we have grown beyond the stage of waiting for technology and instead should be directing it? If so, the first thing we might want to do is seek to understand the factors that promote technological development. Maybe we should look for an answer to the question, "Why does technology happen?"

BIBLIOGRAPHY

Bronowski, Jacob. *The Ascent of Man.* Little Brown and Co., 1974.

Burke, James. *The Axemaker's Gift: A Double-Edged History of Human Culture.* Putnam Publishing Group, 1995.

Feynman, Richard, et al. *What Do You Care What Other People Think? Further Adventures of a Curious Character.* W. W. Norton and Company, 2001.

Graham, Margaret B. W., and Alec T. Shuldner. *Corning and the Craft of Innovation.* Oxford University Press, 2001.

Smith, Cyril S. *A Search for Structure.* MIT Press, 1981.

Wade, Wyn Craig. *The Titanic: End of a Dream.* Penguin USA, 1992.

Much of the research for this book was done via Internet searches, which provided a wealth of information that eventually found its way into this work. The websites of the National Transportation Safety Board, the National Aeronautics and Space Administration, and the Department of Energy were particularly helpful.

INDEX

About the Author

Dr. Mark E. Eberhart lives in Denver and is a professor of chemistry and materials science at the Colorado School of Mines in Golden, Colorado. He balances his time between research to better understand the nature of the chemical bond, teaching, restoring beautiful old homes to new grandeur, cycling, and weekend commutes to Santa Fe, New Mexico, where his wife, Cheryl, practices law.